SUPERSYMMETRIC SOLITONS

In the last decade methods and techniques based on supersymmetry have provided deep insights in quantum chromodynamics and other non-supersymmetric gauge theories at strong coupling. This book summarizes major advances in critical solitons in supersymmetric theories, and their implications for understanding basic dynamical regularities of non-supersymmetric theories.

After an extended introduction on the theory of critical solitons, including a historical introduction, the authors focus on three topics: non-Abelian strings and confined monopoles; reducing the level of supersymmetry; and domain walls as D brane prototypes. They also provide a thorough review of issues at the cutting edge, such as non-Abelian flux tubes. The book presents an extensive summary of the current literature so that researchers in this field can understand the background and related issues. This title, first published in 2009, has been reissued as an Open Access publication on Cambridge Core.

MIKHAIL SHIFMAN is the Ida Cohen Fine Professor of Physics at the University of Minnesota, and is one of the world leading experts on quantum chromodynamics and non-perturbative supersymmetry. In 1999 he received the Sakurai Prize for Theoretical Particle Physics, and in 2006 he was awarded the Julius Edgar Lilienfeld Prize for outstanding contributions to physics. He is the author of several books, over 300 scientific publications, and a number of popular articles and articles on the history of high-energy physics.

ALEXEI YUNG is a Senior Researcher in the Theoretical Department at the Petersburg Nuclear Physics Institute, Russia, and a Visiting Professor at the William I. Fine Theoretical Physics Institute. His research interests lie in non-perturbative dynamics of non-Abelian supersymmetric gauge theories and its interplay with string theory, and the problem of color confinement in non-Abelian gauge theories. Many of his recent advances in these areas are included in this book.

CAMBRIDGE MONOGRAPHS ON MATHEMATICAL PHYSICS

General Editors: P. V. Landshoff, D. R. Nelson, S. Weinberg

L. O'Raifeartaigh *Group Structure of Gauge Theories*[†]

T. Ortín *Gravity and Strings*[†]

A. M. Ozorio de Almeida *Hamiltonian Systems: Chaos and Quantization*[†]

R. Penrose and W. Rindler *Spinors and Space-Time Volume 1: Two-Spinor Calculus and Relativistic Fields*[†]

R. Penrose and W. Rindler *Spinors and Space-Time Volume 2: Spinor and Twistor Methods in Space-Time Geometry*[†]

S. Pokorski *Gauge Field Theories, 2nd edition*[†]

J. Polchinski *String Theory Volume 1: An Introduction to the Bosonic String*

J. Polchinski *String Theory Volume 2: Superstring Theory and Beyond*

V. N. Popov *Functional Integrals and Collective Excitations*[†]

R. J. Rivers *Path Integral Methods in Quantum Field Theory*[†]

R. G. Roberts *The Structure of the Proton: Deep Inelastic Scattering*[†]

C. Rovelli *Quantum Gravity*[†]

W. C. Saslaw *Gravitational Physics of Stellar and Galactic Systems*[†]

M. Shifman and A. Yung *Supersymmetric Solitons*

H. Stephani, D. Kramer, M. MacCallum, C. Hoenselaers and E. Herlt *Exact Solutions of Einstein's Field Equations, 2nd edition*

J. Stewart *Advanced General Relativity*[†]

T. Thiemann *Modern Canonical Quantum General Relativity*

D. J. Toms *The Schwinger Action Principle and Effective Action*

A. Vilenkin and E. P. S. Shellard *Cosmic Strings and Other Topological Defects*[†]

R. S. Ward and R. O. Wells, Jr *Twistor Geometry and Field Theory*[†]

J. R. Wilson and G. J. Mathews *Relativistic Numerical Hydrodynamics*

[†] Issued as a paperback

Supersymmetric Solitons

M. SHIFMAN

William I. Fine Theoretical Physics Institute
University of Minnesota

A. YUNG

William I. Fine Theoretical Physics Institute
University of Minnesota
Petersburg Nuclear Physics Institute
Institute of Theoretical and Experimental Physics

CAMBRIDGE
UNIVERSITY PRESS

Shaftesbury Road, Cambridge CB2 8EA, United Kingdom

One Liberty Plaza, 20th Floor, New York, NY 10006, USA

477 Williamstown Road, Port Melbourne, VIC 3207, Australia

314–321, 3rd Floor, Plot 3, Splendor Forum, Jasola District Centre, New Delhi – 110025, India

103 Penang Road, #05-06/07, Visioncrest Commercial, Singapore 238467

Cambridge University Press is part of Cambridge University Press & Assessment,
a department of the University of Cambridge.

We share the University's mission to contribute to society through the pursuit of
education, learning and research at the highest international levels of excellence.

www.cambridge.org
Information on this title: www.cambridge.org/9781009402170

DOI: 10.1017/9781009402200

First published 2009
Reissued as OA 2023

A catalogue record for this publication is available from the British Library.

ISBN 978-1-009-40217-0 Hardback
ISBN 978-1-009-40222-4 Paperback

Caricature of Alyosha Yung and Misha Shifman by Andrey Feldshteyn, 2007.

Contents

II LONG JOURNEY

Acknowledgments

We are grateful to Adam Ritz, David Tong, and Arkady Vainshtein for useful discussions.

The work of M.S. was supported in part by DOE grant DE-FG02-94ER408. The work of A.Y. was supported by FTPI, University of Minnesota, by RFBR Grant No. 06-02-16364a and by Russian State Grant for Scientific School RSGSS-11242003.2.

Abbreviations

AdS   Anti de Sitter
ANO   Abrikosov–Nielsen–Olesen
BPS   Bogomol'nyi–Prasad–Sommerfield
CC   Central Charge
CMS   Curve(s) of the Marginal Stability
CFT   Conformal Field Theory
CFIV   Cecotti–Fendley–Intriligator–Vafa
FI   Fayet–Iliopoulos
IR   Infrared
NSVZ   Navikov–Shifman–Vainshtein–Zakharov
QCD   Quantum Chromodynamics
SUSY   Supersymmetry, Supersymmetric
SQCD   Supersymmetric Quantum Chromodynamics
SQED   Supersymmetric Quantum Electrodynamics
UV   Ultraviolet
VEV   Vacuum Expectation Value

1

Introduction

It is well known that supersymmetric theories may have Bogomol'nyi–Prasad–Sommerfield (BPS) sectors in which some data can be computed at strong coupling even when the full theory is not solvable. Historically, this is how the first exact results on particle spectra were obtained [1]. Seiberg–Witten's breakthrough results [2, 3] in the mid 1990s gave an additional motivation to the studies of the BPS sectors.

BPS solitons can emerge in those supersymmetric theories in which superalgebras are centrally extended. In many instances the corresponding central charges are seen at the classical level. In some interesting models central charges appear as quantum anomalies.

First studies of BPS solitons (sometimes referred to as critical solitons) in supersymmetric theories at weak coupling date back to the 1970s. De Vega and Schaposnik were the first to point out [4] that a model in which classical equations of motion can be reduced to first-order Bogomol'nyi–Prasad–Sommerfeld (BPS) equations [5, 6] is, in fact, a bosonic reduction of a supersymmetric theory. Already in 1977 critical soliton solutions were obtained in the superfield form in some two-dimensional models [7]. In the same year miraculous cancellations occurring in calculations of quantum corrections to soliton masses were noted in [8] (see also [9]). It was observed that for BPS solitons the boson and fermion modes are degenerate and their number is balanced. It was believed (incorrectly, we hasten to add) that the soliton masses receive no quantum corrections. The modern – correct – version of this statement is as follows: if a soliton is BPS-saturated at the classical level and belongs to a shortened supermultiplet, it stays BPS-saturated after quantum corrections, and its mass exactly coincides with the central charge it saturates. The latter may or may not be renormalized. Often – but not always – central charges that do not vanish at the classical level and have quantum anomalies *are* renormalized. Those that emerge as anomalies and have no classical part typically receive no renormalizations. In many instances holomorphy protects central charges against renormalizations.

Critical solitons play a special role in gauge field theories. Numerous parallels between such solitonic objects and basic elements of string theory have been revealed in recent years. At first, the relation between string theory and supersymmetric gauge theories was mostly a "one-way street" – from strings to field theory. Now it is becoming exceedingly more evident that field-theoretic methods and results, in their turn, provide insights in string theory.

String theory, which emerged from dual hadronic models in the late 1960s and 70s, elevated to the "theory of everything" in the 1980s and 90s when it experienced an unprecedented expansion, has seemingly entered a "return-to-roots" stage. The task of finding solutions to "down-to-earth" problems of QCD and other gauge theories by using results and techniques of string/D-brane theory is currently recognized by many as one of the most important and exciting goals of the community. In this area the internal logic of development of string theory is fertilized by insights and hints obtained from field theory. In fact, this is a very healthy process of cross-fertilization.

If supersymmetric gauge theories are, in a sense, dual to string/D-brane theory – as is generally believed to be the case – they must support domain walls (of the D-brane type) [10], and we know, they do [11, 12]. A D-brane is defined as a hypersurface on which a string may end. In field theory both the brane and the string arise as BPS solitons, the brane as a domain wall and the string as a flux tube. If their properties reflect those inherent to string theory, at least to an extent, the flux tube must end on the wall. Moreover, the wall must house gauge fields living on its world volume, under which the end of the string is charged.

The purpose of this review is to summarize developments in critical solitons in two, three and four dimensions, with emphasis on four dimensions and on most recent results. A large variety of BPS-saturated solitons exist in four-dimensional field theories: domain walls, flux tubes (strings), monopoles and dyons, and various junctions of the above objects. A list of recent discoveries includes localization of gauge fields on domain walls, non-Abelian strings that can end on domain walls, developed boojums, confined monopoles attached to strings, and other remarkable findings. The BPS nature of these objects allows one to obtain a number of exact results. In many instances nontrivial dynamics of the bulk theories we will consider lead to effective low-energy theories in the world volumes of domain walls and strings (they are related to zero modes) exhibiting novel dynamical features that are interesting by themselves.

We do not try to review the vast literature accumulated since the mid 1990s in its entirety. A comparison with a huge country the exploration of which is not yet completed is in order here. Instead, we suggest what may be called "travel diaries" of the participants of the exploratory expedition. Recent publications [13, 14, 15, 16, 17] facilitate our task since they present the current developments in this field from a complementary point of view.

The "diaries" are organized in two parts. The first part (entitled "Short excursion") is a bird's eye view of the territory. It gives a brief and largely nontechnical introduction to basic ideas lying behind supersymmetric solitons and particular applications. It is designed in such a way as to present a general perspective that would be understandable to anyone with an elementary knowledge in classical and quantum fields, and supersymmetry.

Here we present some historic remarks, catalog relevant centrally extended superalgebras and review basic building blocks we consistently deal with – domain walls, flux tubes, and monopoles – in their classic form. The word "classic" is used here not in the meaning "before quantization" but, rather, in the meaning "recognized and cherished in the community for years."

The second part (entitled "Long journey") is built on other principles. It is intended for those who would like to delve in this subject thoroughly, with its specific methods and technical devices. We put special emphasis on recent developments having direct relevance to QCD and gauge theories at large, such as non-Abelian flux tubes (strings), non-Abelian monopoles confined on these strings, gauge field localization on domain walls, etc. We start from presenting our benchmark model, which has extended $\mathcal{N} = 2$ supersymmetry. Here we go well beyond conceptual foundations, investing efforts in detailed discussions of particular problems and aspects of our choosing. Naturally, we choose those problems and aspects which are instrumental in the novel phenomena mentioned above. In addition to walls, strings and monopoles, we also dwell on the string-wall junctions which play a special role in the context of dualization.

Our subsequent logic is from $\mathcal{N} = 2$ to $\mathcal{N} = 1$ and further on. Indeed, in certain instances we are able to descend to non-supersymmetric gauge theories which are very close relatives of QCD. In particular, we present a fully controllable weakly coupled model of the Meissner effect which exhibits quite nontrivial (strongly coupled) dynamics on the string world sheet. One can draw direct parallels between this consideration and the issue of k-strings in QCD.

Part I

Short excursion

2

Central charges in superalgebras

In this Section we will briefly review general issues related to central charges (CC) in superalgebras.

2.1 History

The first superalgebra in four-dimensional field theory was derived by Golfand and Likhtman [18] in the form

$$\{\bar{Q}_{\dot{\alpha}} Q_\beta\} = 2P_\mu \left(\sigma^\mu\right)_{\alpha\beta}, \quad \{\bar{Q}_\alpha \bar{Q}_\beta\} = \{Q_\alpha Q_\beta\} = 0, \qquad (2.1.1)$$

i.e. with no central charges. Possible occurrence of CC (elements of superalgebra commuting with all other operators) was first mentioned in an unpublished paper of Lopuszanski and Sohnius [19] where the last two anticommutators were modified as

$$\{Q_\alpha^I Q_\beta^G\} = Z_{\alpha\beta}^{IG}. \qquad (2.1.2)$$

The superscripts I, G mark extended supersymmetry. A more complete description of superalgebras with CC in quantum field theory was worked out in [20]. The only central charges analyzed in this paper were Lorentz scalars (in four dimensions), $Z_{\alpha\beta} \sim \varepsilon_{\alpha\beta}$. Thus, by construction, they could be relevant only to extended supersymmetries.

A few years later, Witten and Olive [1] showed that in supersymmetric theories with solitons, central extension of superalgebras is typical; topological quantum numbers play the role of central charges.

It was generally understood that superalgebras with (Lorentz-scalar) central charges can be obtained from superalgebras without central charges in higher-dimensional space-time by interpreting some of the extra components of the momentum as CC's (see e.g. [21]). When one compactifies extra dimensions one obtains an extended supersymmetry; the extra components of the momentum act as scalar central charges.

Algebraic analysis extending that of [20] carried out in the early 1980s (see e.g. [22]) indicated that the super-Poincaré algebra admits CC's of a more general form, but the dynamical role of additional tensorial charges was not recognized until much later. Now it is common knowledge that central charges that originate from operators other than the energy-momentum operator in higher dimensions can play a crucial role. These tensorial central charges take non-vanishing values on extended objects such as strings and membranes.

Central charges that are antisymmetric tensors in various dimensions were introduced (in the supergravity context, in the presence of p-branes) in Ref. [23] (see also [24, 25]). These CC's are relevant to extended objects of the domain wall type (membranes). Their occurrence in four-dimensional super-Yang–Mills theory (as a quantum anomaly) was first observed in [11]. A general theory of central extensions of superalgebras in three and four dimensions was discussed in Ref. [26]. It is worth noting that those central charges that have the Lorentz structure of Lorentz vectors were not considered in [26]. The gap was closed in [27].

2.2 Minimal supersymmetry

The minimal number of supercharges ν_Q in various dimensions is given in Table 2.1. Two-dimensional theories with a single supercharge, although algebraically possible, are quite exotic. In "conventional" models in $D = 2$ with local interactions the minimal number of supercharges is two.

The minimal number of supercharges in Table 2.1 is given for a real representation. Then, it is clear that, generally speaking, the maximal possible number of CC's is determined by the dimension of the symmetric matrix $\{Q_i Q_j\}$ of the size $\nu_Q \times \nu_Q$, namely,

$$\nu_{CC} = \frac{\nu_Q(\nu_Q + 1)}{2}. \tag{2.2.1}$$

In fact, D anticommutators have the Lorentz structure of the energy-momentum operator P_μ. Therefore, up to D central charges could be absorbed in P_μ, generally speaking. In particular situations this number can be smaller, since although algebraically the corresponding CC's have the same structure as P_μ, they are dynamically distinguishable. The point is that P_μ is uniquely defined through the conserved and symmetric energy-momentum tensor of the theory.

Additional dynamical and symmetry constraints can further diminish the number of independent central charges, see e.g. Section 2.2.1.

The total set of CC's can be arranged by classifying CC's with respect to their Lorentz structure. Below we will present this classification for $D = 2, 3$ and 4, with

Table 2.1. *The minimal number of supercharges, the complex dimension of the spinorial representation and the number of additional conditions (i.e. the Majorana and/or Weyl conditions).*

D	2	3	4	5	6	7	8	9	10
v_Q	$(1^*)\,2$	2	4	8	8	8	16	16	16
$\mathrm{Dim}(\psi)_C$	2	2	4	4	8	8	16	16	32
# cond.	2	1	1	0	1	1	1	1	2

special emphasis on the four-dimensional case. In Section 2.3 we will deal with $\mathcal{N} = 2$ superalgebras.

2.2.1 $D = 2$

Consider two-dimensional non-chiral theories with two supercharges. From the discussion above, on purely algebraic grounds, three CC's are possible: one Lorentz-scalar and a two-component vector,

$$\{Q_\alpha, Q_\beta\} = 2(\gamma^\mu \gamma^0)_{\alpha\beta}(P_\mu + Z_\mu) + i(\gamma^5 \gamma_0)_{\alpha\beta} Z. \tag{2.2.2}$$

We refer to Appendix A for our conventions regarding gamma matrices. $Z^\mu \neq 0$ would require existence of a vector order parameter taking distinct values in different vacua. Indeed, if this central charge existed, its current would have the form

$$\zeta_\nu^\mu = \varepsilon_{\nu\rho}\, \partial^\rho A^\mu, \quad Z^\mu = \int \zeta_0^\mu \, dz,$$

where A^μ is the above-mentioned order parameter. However, $\langle A^\mu \rangle \neq 0$ will break Lorentz invariance and supersymmetry of the vacuum state. This option will not be considered. Limiting ourselves to supersymmetric vacua we conclude that a single (real) Lorentz-scalar central charge Z is possible in $\mathcal{N} = 1$ theories. This central charge is saturated by kinks.

2.2.2 $D = 3$

The central charge allowed in this case is a Lorentz-vector Z_μ, i.e.

$$\{Q_\alpha, Q_\beta\} = 2(\gamma^\mu \gamma^0)_{\alpha\beta}(P_\mu + Z_\mu). \tag{2.2.3}$$

One should arrange Z_μ to be orthogonal to P_μ. In fact, this is the scalar central charge of Section 2.2.1 elevated by one dimension. Its topological current can be written as

$$\zeta_{\mu\nu} = \varepsilon_{\mu\nu\rho}\,\partial^\rho A, \quad Z_\mu = \int d^2x\,\zeta_{\mu 0}. \tag{2.2.4}$$

By an appropriate choice of the reference frame Z_μ can always be reduced to a real number times $(0,0,1)$. This central charge is associated with a domain line oriented along the second axis.

Although from the general relation (2.2.3) it is pretty clear why BPS vortices cannot appear in theories with two supercharges, it is instructive to discuss this question from a slightly different standpoint. Vortices in three-dimensional theories are localized objects, particles (BPS vortices in $2+1$ dimensions were previously considered in [28]; see also references therein). The number of broken translational generators is d, where d is the soliton's co-dimension, $d = 2$ in the case at hand. Then *at least* d supercharges are broken. Since we have only two supercharges in the problem at hand, both must be broken. This simple argument tells us that for a 1/2-BPS vortex the minimal matching between bosonic and fermionic zero modes in the (super) translational sector is one-to-one.

Consider now a putative BPS vortex in a theory with minimal $\mathcal{N} = 1$ supersymmetry (SUSY) in $2+1$D. Such a configuration would require a world volume description with two bosonic zero modes, but only one fermionic mode. This is not permitted by the argument above, and indeed no configurations of this type are known. Vortices always exhibit at least two fermionic zero modes and can be BPS-saturated only in $\mathcal{N} = 2$ theories.

2.2.3 D = 4

Maximally one can have 10 CC's which are decomposed into Lorentz representations as $(0,1) + (1,0) + (1/2,1/2)$:

$$\{Q_\alpha, \bar{Q}_{\dot\alpha}\} = 2(\gamma^\mu)_{\alpha\dot\alpha}(P_\mu + Z_\mu), \tag{2.2.5}$$

$$\{Q_\alpha, Q_\beta\} = (\Sigma^{\mu\nu})_{\alpha\beta} Z_{[\mu\nu]}, \tag{2.2.6}$$

$$\{\bar{Q}_{\dot\alpha}, \bar{Q}_{\dot\beta}\} = (\bar\Sigma^{\mu\nu})_{\dot\alpha\dot\beta} \bar{Z}_{[\mu\nu]}, \tag{2.2.7}$$

where $(\Sigma^{\mu\nu})_{\alpha\beta} = (\sigma^\mu)_{\alpha\dot\alpha}(\bar\sigma^\nu)^{\dot\alpha}_\beta$ is a chiral version of $\sigma^{\mu\nu}$ (see e.g. [29]). The antisymmetric tensors $Z_{[\mu\nu]}$ and $\bar{Z}_{[\mu\nu]}$ are associated with domain walls, and reduce to a complex number and a spatial vector orthogonal to the domain wall. The $(1/2, 1/2)$ CC Z_μ is a Lorentz vector orthogonal to P_μ. It is associated with strings (flux tubes), and reduces to one real number and a three-dimensional unit spatial vector parallel to the string.

2.3 Extended SUSY

In four dimensions one can extend superalgebra up to $\mathcal{N} = 4$, which corresponds
to sixteen supercharges. Reducing this to lower dimensions we get a rich variety
of extended superalgebras in $D = 3$ and 2. In fact, in two dimensions the Lorentz
invariance provides a much weaker constraint than in higher dimensions, and one
can consider a wider set of (p, q) superalgebras comprising $p + q = 4$, 8, or 16
supercharges. We will not pursue a general solution; instead, we will limit our task
to (i) analysis of central charges in $\mathcal{N} = 2$ in four dimensions; (ii) reduction of
the minimal SUSY algebra in $D = 4$ to $D = 2$ and 3, namely the $\mathcal{N} = 2$ SUSY
algebra in those dimensions. Thus, in two dimensions we will consider only the
non-chiral $\mathcal{N} = (2, 2)$ case. As should be clear from the discussion above, in the
dimensional reduction the maximal number of CC's stays intact. What changes is
the decomposition in Lorentz and R-symmetry irreducible representations.

2.3.1 $\mathcal{N} = 2$ in $D = 2$

Let us focus on the non-chiral $\mathcal{N} = (2, 2)$ case corresponding to dimensional
reduction of the $\mathcal{N} = 1$, $D = 4$ algebra. The tensorial decomposition is as follows:

$$\{Q_\alpha^I, Q_\beta^J\} = 2(\gamma^\mu \gamma^0)_{\alpha\beta} \left[(P_\mu + Z_\mu)\delta^{IJ} + Z_\mu^{(IJ)} \right] + 2i\,(\gamma^5\gamma^0)_{\alpha\beta}\, Z^{\{IJ\}}$$
$$+ 2i\,\gamma_{\alpha\beta}^0 Z^{[IJ]}, \quad I, J = 1, 2. \tag{2.3.1}$$

Here $Z^{[IJ]}$ is antisymmetric in I, J; $Z^{\{IJ\}}$ is symmetric while $Z^{(IJ)}$ is symmetric
and traceless. We can discard all vectorial central charges Z_μ^{IJ} for the same reasons
as in Section 2.2.1. Then we are left with two Lorentz singlets $Z^{(IJ)}$, which represent
the reduction of the domain wall charges in $D = 4$ and two Lorentz singlets $\mathrm{Tr}\, Z^{\{IJ\}}$
and $Z^{[IJ]}$, arising from P_2 and the vortex charge in $D = 3$ (see Section 2.3.2). These
central charges are saturated by kinks.

Summarizing, the $(2, 2)$ superalgebra in $D = 2$ is

$$\{Q_\alpha^I, Q_\beta^J\} = 2(\gamma^\mu \gamma^0)_{\alpha\beta}\, P_\mu\, \delta^{IJ} + 2i(\gamma^5\gamma^0)_{\alpha\beta}\, Z^{\{IJ\}} + 2i\,\gamma_{\alpha\beta}^0 Z^{[IJ]}. \tag{2.3.2}$$

It is instructive to rewrite Eq. (2.3.2) in terms of complex supercharges Q_α and $Q_\beta^\dagger$
corresponding to four-dimensional Q_α, $\bar{Q}_{\dot\alpha}$, see Section 2.2.3. Then

$$\{Q_\alpha, Q_\beta^\dagger\}(\gamma^0)_{\beta\gamma} = 2\left[P_\mu \gamma^\mu + Z\frac{1 - \gamma_5}{2} + Z^\dagger\frac{1 + \gamma_5}{2} \right]_{\alpha\gamma}, \tag{2.3.3}$$

$$\{Q_\alpha, Q_\beta\}(\gamma^0)_{\beta\gamma} = -2Z'\,(\gamma_5)_{\alpha\gamma}, \quad \{Q_\alpha^\dagger, Q_\beta^\dagger\}(\gamma^0)_{\beta\gamma} = 2Z'^\dagger\,(\gamma_5)_{\alpha\gamma}.$$

The algebra contains two complex central charges, Z and Z'. In terms of components $Q_\alpha = (Q_R, Q_L)$ the nonvanishing anticommutators are

$$\{Q_L, Q_L^\dagger\} = 2(H + P), \quad \{Q_R, Q_R^\dagger\} = 2(H - P),$$
$$\{Q_L, Q_R^\dagger\} = 2i\, Z, \quad\quad\ \{Q_R, Q_L^\dagger\} = -2i\, Z^\dagger, \quad\quad (2.3.4)$$
$$\{Q_L, Q_R\} = 2i\, Z', \quad\quad\ \{Q_R^\dagger, Q_L^\dagger\} = -2i\, Z'^\dagger.$$

It exhibits the automorphism $Q_R \leftrightarrow Q_R^\dagger$, $Z \leftrightarrow Z'$ associated [30] with the transition to a mirror representation [31]. The complex central charges Z and Z' can be readily expressed in terms of real $Z^{\{IJ\}}$ and $Z^{[IJ]}$,

$$Z = Z^{[12]} + \frac{i}{2}\left(Z^{\{11\}} + Z^{\{22\}}\right), \quad Z' = \frac{Z^{\{12\}} + Z^{\{21\}}}{2} - i\,\frac{Z^{\{11\}} - Z^{\{22\}}}{2}. \quad (2.3.5)$$

Typically, in a given model either Z or Z' vanish. A practically important example to which we will repeatedly turn below (e.g. Sections 3.5 and 4.5.3) is provided by the so-called twisted-mass-deformed $CP(N-1)$ model [32]. The central charge Z emerges in this model at the classical level. At the quantum level it acquires additional anomalous terms [33, 34]. Both $Z \neq 0$ and $Z' \neq 0$ simultaneously in a contrived model [33] in which the Lorentz symmetry and a part of supersymmetry are spontaneously broken.

2.3.2 $\mathcal{N} = 2$ in $D = 3$

The superalgebra can be decomposed into Lorentz and R-symmetry tensorial structures as follows:

$$\{Q_\alpha^I, Q_\beta^J\} = 2(\gamma^\mu \gamma^0)_{\alpha\beta}[(P_\mu + Z_\mu)\delta^{IJ} + Z_\mu^{(IJ)}] + 2i\,\gamma_{\alpha\beta}^0 Z^{[IJ]}, \quad (2.3.6)$$

where all central charges above are real. The maximal set of 10 CC's enter as a triplet of spacetime vectors Z_μ^{IJ} and a singlet $Z^{[IJ]}$. The singlet CC is associated with vortices (or lumps), and corresponds to the reduction of the $(1/2,1/2)$ charge or the 4th component of the momentum vector in $D = 4$. The triplet Z_μ^{IJ} is decomposed into an R-symmetry singlet Z_μ, algebraically indistinguishable from the momentum, and a traceless symmetric combination $Z_\mu^{(IJ)}$. The former is equivalent to the vectorial charge in the $\mathcal{N} = 1$ algebra, while $Z_\mu^{(IJ)}$ can be reduced to a complex number and vectors specifying the orientation. We see that these are the direct reduction of the $(0,1)$ and $(1,0)$ wall charges in $D = 4$. They are saturated by domain lines.

2.3.3 On extended supersymmetry (eight supercharges) in $D = 4$

Complete algebraic analysis of all tensorial central charges in this problem is analogous to the previous cases and is rather straightforward. With eight supercharges the maximal number of CC's is 36. Dynamical aspect is less developed – only a modest fraction of the above 36 CC's are known to be non-trivially realized in models studied in the literature. We will limit ourselves to a few remarks regarding the well-established CC's. We will use a complex (holomorphic) representation of the supercharges. Then the supercharges are labeled as follows

$$Q_\alpha^F, \quad \bar{Q}_{\dot\alpha G}, \quad \alpha, \dot\alpha = 1, 2, \quad F, G = 1, 2. \tag{2.3.7}$$

On general grounds one can write

$$\{Q_\alpha^F, \bar{Q}_{\dot\alpha G}\} = 2\delta_G^F P_{\alpha\dot\alpha} + 2(Z_G^F)_{\alpha\dot\alpha},$$

$$\{Q_\alpha^F, Q_\beta^G\} = 2Z_{\{\alpha\beta\}}^{\{FG\}} + 2\varepsilon_{\alpha\beta}\,\varepsilon^{FG}\,Z,$$

$$\{\bar{Q}_{\dot\alpha F}, \bar{Q}_{\dot\beta G}\} = 2\left(\bar{Z}_{\{FG\}}\right)_{\{\dot\alpha\dot\beta\}} + 2\varepsilon_{\dot\alpha\dot\beta}\,\varepsilon_{FG}\,\bar{Z}. \tag{2.3.8}$$

Here $(Z_G^F)_{\alpha\dot\alpha}$ are four vectorial central charges (1/2, 1/2), (16 components altogether) while $Z_{\{\alpha\beta\}}^{\{FG\}}$ and the complex conjugate are (1,0) and (0,1) central charges. Since the matrix $Z_{\{\alpha\beta\}}^{\{FG\}}$ is symmetric with respect to F, G, there are three flavor components, while the total number of components residing in (1,0) and (0,1) central charges is 18. Finally, there are two scalar central charges, Z and $\bar{Z}$.

Dynamically the above central charges can be described as follows. The scalar CC's Z and $\bar{Z}$ are saturated by monopoles/dyons. One vectorial central charge Z_μ (with the additional condition $P^\mu Z_\mu = 0$) is saturated [35] by Abrikosov–Nielsen–Olesen string (ANO for short) [36]. A (1,0) central charge with $F = G$ is saturated by domain walls [37].

Let us briefly discuss the Lorentz-scalar central charges in Eq. (2.3.8) that are saturated by monopoles/dyons. They will be referred to as monopole central charges. A rather dramatic story is associated with them. Historically they were the first to be introduced within the framework of an extended 4D superalgebra [19, 20]. On the dynamical side, they appeared as the first example of the "topological charge ↔ central charge" relation revealed by Witten and Olive in their pioneering paper [1]. Twenty years later, the $\mathcal{N} = 2$ model where these central charges first appeared, was solved by Seiberg and Witten [2, 3], and the exact masses of the BPS-saturated monopoles/dyons found. No direct comparison with the operator expression for the central charges was carried out, however. In Ref. [38] it was noted that for the Seiberg–Witten formula to be valid, a boson-term anomaly should exist in the monopole central charges. Even before [38] a fermion-term anomaly was identified [37], which plays a crucial role [39] for the monopoles in the Higgs regime (confined monopoles).

3

The main building blocks

3.1 Domain walls

3.1.1 Preliminaries

In four dimensions domain walls are two-dimensional extended objects. In three dimensions they become domain lines, while in two dimensions they reduce to kinks which can be considered as particles since they are localized. Embeddings of bosonic models supporting kinks in $\mathcal{N} = 1$ supersymmetric models in two dimensions were first discussed in [1, 7]. Occasional remarks on kinks in models with four supercharges of the type of the Wess–Zumino models [40] can be found in the literature in the 1980s but they went unnoticed. The only issue which caused much interest and debate in the 1980s was the issue of quantum corrections to the BPS kink mass in 2D models with $\mathcal{N} = 1$ supersymmetry.

The mass of the BPS saturated kinks in two dimensions must be equal to the central charge Z in Eq. (2.2.2). The simplest two-dimensional model with two supercharges, admitting solitons, was considered in [41]. In components the Lagrangian takes the form

$$\mathcal{L} = \frac{1}{2}\left(\partial_\mu \phi \, \partial^\mu \phi + \bar{\psi} \, i \, \partial\!\!\!/ \psi + F^2\right) + \mathcal{W}'(\phi)F - \frac{1}{2}\mathcal{W}''(\phi)\bar{\psi}\psi, \quad (3.1.1)$$

where ϕ is a real field, ψ is a two-component Majorana spinor in two dimensions, and $\mathcal{W}(\phi)$ is a real "superpotential" which in the simplest case takes the form

$$\mathcal{W}(\phi) = \frac{m^2}{\lambda}\phi - \frac{\lambda}{3}\phi^3. \quad (3.1.2)$$

Moreover, the auxiliary field F can be eliminated by virtue of the classical equation of motion, $F = -\mathcal{W}'$. This is a real reduction (two supercharges) of the

15

Wess–Zumino model (Section 3.1.2). The kink (antikink) BPS equation is

$$\partial_z \phi = \pm \frac{dW}{d\phi}, \tag{3.1.3}$$

with the boundary condition that $\phi(z)$ tends to two distinct vacua, $\phi_{\text{vac}} = \pm m/\lambda$ at $z \to \pm\infty$. It can be readily integrated.

The story of kinks in this model is long and dramatic. In the very beginning it was argued [41] that, due to a residual supersymmetry, the mass of the soliton calculated at the classical level remains intact at the one-loop level. A few years later it was noted [42] that the non-renormalization theorem [41] cannot possibly be correct, since the classical soliton mass is proportional to m^3/λ^2 (where m and λ are the bare mass parameter and coupling constant, respectively), and the physical mass of the scalar field gets a logarithmically infinite renormalization. Since the soliton mass is an observable physical parameter, it must stay finite in the limit $M_{\text{uv}} \to \infty$, where M_{uv} is the ultraviolet cut off. This implies, in turn, that the quantum corrections cannot vanish – they "dress" m in the classical expression, converting the bare mass parameter into the renormalized one. The one-loop renormalization of the soliton mass was first calculated in [42]. Technically the emergence of the one-loop correction was attributed to a "difference in the density of states in continuum in the boson and fermion operators in the soliton background field." The subsequent work [43] dealt with the renormalization of the central charge, with the conclusion that the central charge is renormalized in just the same way as the kink mass, so that the saturation condition is not violated.

Then many authors repeated one-loop calculations for the kink mass and/or central charge [44, 45, 46, 47, 48, 49, 50, 51, 52, 53, 54]. The results reported and the conclusion of saturation/non-saturation oscillated with time, with little sign of convergence. Needless to say, all authors agreed that the logarithmically divergent term in Z matched the renormalization of m. However, the finite (non-logarithmic) term varied from work to work, sometimes even in the successive works of the same authors. Polemics continued unabated through the 1990s. For instance, Nastase *et al.* [53], presenting a perfectly valid calculation of the kink mass, concluded that the BPS saturation was violated at one loop. This assertion reversed the earlier trend [42, 49, 50], according to which the kink mass and the corresponding central charge are renormalized in a concerted way. A somewhat later publication [54] again changed the scene, advocating BPS saturation. However, a dimensionally regularized kink mass determined in [54] was not consistent with that found in [53].

The story culminated in 1998 with the discovery of a quantum anomaly in the central charge [55]. Classically, the kink central charge Z is equal to the difference between the values of the superpotential $\mathcal{W}$ at spatial infinities,

$$Z = \mathcal{W}[\phi(z = \infty)] - \mathcal{W}[\phi(z = -\infty)]. \tag{3.1.4}$$

This is known from the pioneering paper [1]. Due to the anomaly, the central charge gets modified in the following way

$$\mathcal{W} \longrightarrow \mathcal{W} + \frac{\mathcal{W}''}{4\pi}, \tag{3.1.5}$$

where the term proportional to $\mathcal{W}''$ is anomalous [55]. The right-hand side of Eq. (3.1.5) must be substituted in the expression for the central charge (3.1.4) instead of $\mathcal{W}$. Inclusion of the additional anomalous term restores the equality between the kink mass and its central charge. The BPS nature is preserved, which is correlated with the fact that the kink supermultiplet is short in the case at hand [56]. All subsequent investigations confirmed this conclusion (see e.g. the review paper [57] and original papers [58] by van Nieuwenhuizen and collaborators).

Critical domain walls in $\mathcal{N} = 1$ four-dimensional theories (four supercharges) started attracting attention in the 1990s. What is the domain wall? It is a two-dimensional object of co-dimension one. It is a field configuration interpolating between vacuum i and vacuum f with some transition domain in the middle. Say, to the left you have vacuum i, to the right you have vacuum f, in the middle you have a transition domain which, for obvious reasons, is referred to as the wall (Fig. 3.1). The most popular model of this time supporting such domain walls was the generalized Wess–Zumino model with the Lagrangian

$$\mathcal{L} = \int d^2\theta \, d^2\bar{\theta} \, K(\bar{\Phi}_a, \Phi_a) + \left(\int d^2\theta \, \mathcal{W}(\Phi) + \text{H.c.} \right) \tag{3.1.6}$$

where K is the Kähler potential and Φ_a stands for a set of the chiral superfields. The number of the chiral superfields can be arbitrary, but the superpotential $\mathcal{W}$ must have at least two critical points, two vacua.

(This model can be considered, upon dimensional reduction, in two dimensions as well.) A popular choice was a trivial Kähler potential,

$$K = \sum_a \bar{\Phi}_a \Phi_a.$$

BPS walls in this system satisfy the first-order differential equations [59, 24, 60, 61, 62]

$$g_{\bar{a}b} \, \partial_z \Phi^b = e^{i\eta} \, \partial_{\bar{a}} \bar{\mathcal{W}}, \tag{3.1.7}$$

where the Kähler metric is given by

$$g_{\bar{a}b} = \frac{\partial^2 K}{\partial \bar{\Phi}^{\bar{a}}\, \partial \Phi^b} \equiv \partial_{\bar{a}} \partial_b K, \tag{3.1.8}$$

and η is the phase of the (1,0) central charge Z as defined in (2.2.6). The phase η depends on the choice of the vacua between which the given domain wall interpolates,

$$Z = 2 \left(\mathcal{W}_{\mathrm{vac}_f} - \mathcal{W}_{\mathrm{vac}_i} \right). \tag{3.1.9}$$

A useful consequence of the BPS equations is that

$$\partial_z \mathcal{W} = e^{i\eta}\, \|\partial_a \mathcal{W}\|^2, \tag{3.1.10}$$

and thus the domain wall describes a straight line in the $\mathcal{W}$-plane connecting the two vacua. Needless to say, the first-order BPS equation (3.1.7) guarantees the validity of the second-order equation of motion. The opposite is not true, generally speaking. However, if one deals with a *single* chiral field Φ, one can prove [63] that the BPS equation does follow from the second-order equation of motion.

Construction and analysis of BPS saturated domain walls in four dimensions crucially depends on the realization of the fact that the central charges relevant to critical domain walls are not Lorentz scalars; rather they transform as $(1,0) + (0,1)$ under the Lorentz transformations. It was a textbook statement ascending to the pioneering paper [20] that $\mathcal{N} = 1$ superalgebras in four dimensions leave place to no central charges. This statement is correct only with respect to Lorentz-scalar central charges. Townsend was the first to note [64] that "supersymmetric branes," being BPS saturated, require the existence of tensorial central charges antisymmetric in the vectorial Lorentz indices. That the anticommutator $\{Q_\alpha, Q_\beta\}$ in four-dimensional Wess–Zumino model contains the (1,0) central charge is obvious. This anticommutator vanishes, however, in super-Yang–Mills theory at the classical level (Section 3.1.3).

3.1.2 Domain wall in the minimal Wess–Zumino model

The Wess–Zumino model describes interactions of an arbitrary number of the chiral superfields. We will consider the minimal Wess–Zumino model [65] which describes one chiral superfield,

$$\Phi(x_L, \theta) = \phi(x_L) + \sqrt{2}\,\theta^\alpha \psi_\alpha(x_L) + \theta^2 F(x_L), \tag{3.1.11}$$

$$(x_L)_{\alpha\dot{\alpha}} = x_{\alpha\dot{\alpha}} \mp 2i\,\theta_\alpha \bar{\theta}_{\dot{\alpha}}, \tag{3.1.12}$$

with the canonical kinetic term $K = \bar{\Phi}\Phi$. In components the Lagrangian has the form

$$\mathcal{L} = (\partial^\mu \bar{\phi})(\partial_\mu \phi) + \psi^\alpha i \partial_{\alpha\dot\alpha} \bar{\psi}^{\dot\alpha} + \bar{F}F + \left\{ F \mathcal{W}'(\phi) - \frac{1}{2}\mathcal{W}''(\phi)\psi^2 + \text{H.c.} \right\}.$$

(3.1.13)

From Eq. (3.1.13) it is obvious that F can be eliminated by virtue of the classical equation of motion,

$$\bar{F} = -\frac{\partial \mathcal{W}(\phi)}{\partial \phi},$$

(3.1.14)

so that the scalar potential describing self-interaction of the field ϕ is

$$V(\phi, \bar{\phi}) = \left| \frac{\partial \mathcal{W}(\phi)}{\partial \phi} \right|^2.$$

(3.1.15)

In what follows we will often denote the chiral superfield and its lowest (bosonic) component by one and the same letter, making no distinction between capital and small ϕ. Usually it is clear from the context what is meant in each particular case.

If one limits oneself to renormalizable theories, the superpotential $\mathcal{W}$ must be a polynomial function of Φ of power not higher than three. In the model at hand, with one chiral superfield, the generic superpotential can be always reduced to the following "standard" form:

$$\mathcal{W}(\Phi) = \frac{m^2}{\lambda}\Phi - \frac{\lambda}{3}\Phi^3.$$

(3.1.16)

The quadratic term can be always eliminated by a redefinition of the field Φ. Moreover, by using symmetries of the model one can always choose the phases of the constants m and λ at will.

The superpotential (3.1.16) implies two degenerate classical vacua,

$$\phi_{\text{vac}} = \pm\frac{m}{\lambda}.$$

(3.1.17)

Both vacua are physically equivalent. This equivalence could be explained by the spontaneous breaking of Z_2 symmetry, $\Phi \to -\Phi$, present in the action.

Field configurations interpolating between two degenerate vacua are the domain walls. They have the following properties: (i) the corresponding solutions are static and depend only on one spatial coordinate; (ii) they are topologically stable and indestructible – once a wall is created it cannot disappear. Assume for definiteness that the wall lies in the xy plane. This is the geometry we will always keep in mind.

Then the wall solution ϕ_w will depend only on z. Since the wall extends indefinitely in the xy plane, its energy E_w is infinite. However, the wall tension T_w (the energy per unit area $T_w = E_w/A$) is finite, in principle measurable, and has a clear-cut physical meaning.

The wall solution of the classical equations of motion superficially looks very similar to that of the two-dimensional kink,

$$\phi_w = \frac{m}{\lambda} \tanh(|m|z). \qquad (3.1.18)$$

Note, however, that the parameters m and λ are not necessarily assumed to be real; the field ϕ is complex in the Wess–Zumino model. A remarkable feature of this solution is that it preserves 1/2 of supersymmetry, much in the same way as the kink of Section 3.1.1. The difference is that 1/2 BPS in the two-dimensional model meant one supercharge, now it means two supercharges.

The SUSY transformations generate the following transformation of the fields:

$$\delta\phi = \sqrt{2}\varepsilon\psi, \quad \delta\psi^\alpha = \sqrt{2}\left[\varepsilon^\alpha F + i\,\partial_\mu\phi\,(\sigma^\mu)^{\alpha\dot{\alpha}}\,\bar{\varepsilon}_{\dot{\alpha}}\right]. \qquad (3.1.19)$$

The domain wall we consider is purely bosonic, $\psi = 0$. Moreover, the BPS equation is

$$F\,|_{\bar{\phi}=\phi_w^*} = -e^{-i\eta}\,\partial_z\phi_w(z), \qquad (3.1.20)$$

where

$$\eta = \arg\frac{m^3}{\lambda^2}, \qquad (3.1.21)$$

and $F = -\partial\bar{\mathcal{W}}/\partial\bar{\phi}$. This is a first-order differential equation. The solution quoted above satisfies this condition. The reason for the occurrence of the phase factor $\exp(-i\eta)$ on the right-hand side of Eq. (3.1.20) will become clear shortly. Note that no analog of this phase factor exists in the two-dimensional $\mathcal{N} = 1$ problem on which we dwelled in Section 3.1.1. There was only a sign ambiguity: two possible choices of signs corresponded to kink versus antikink.

If the BPS equation is satisfied, then the second supertransformation in Eq. (3.1.19) reduces to

$$\delta\psi_\alpha \propto \varepsilon_\alpha + i\,e^{i\eta}\,(\sigma^z)_{\alpha\dot{\alpha}}\,\bar{\varepsilon}^{\dot{\alpha}}. \qquad (3.1.22)$$

The right-hand side vanishes provided that

$$\varepsilon_\alpha = -i\,e^{i\eta}\,(\sigma^z)_{\alpha\dot{\alpha}}\,\bar{\varepsilon}^{\dot{\alpha}}. \qquad (3.1.23)$$

This picks up two supertransformations (out of four) which do not act on the domain wall (alternatively people often say that they act trivially). *Quod erat demonstrandum.*

Now, let us calculate the wall tension. To this end we rewrite the expression for the energy functional as

$$\mathcal{E} = \int_{-\infty}^{+\infty} dz \left[\partial_z \bar{\phi} \, \partial_z \phi + \bar{F} F \right]$$

$$\equiv \int_{-\infty}^{+\infty} dz \left\{ \left[e^{-i\eta} \partial_z \mathcal{W} + \text{H.c.} \right] + \left| \partial_z \phi + e^{i\eta} F \right|^2 \right\}, \qquad (3.1.24)$$

where ϕ is assumed to depend only on z. In the literature this procedure is called the Bogomol'nyi completion. The second term on the right-hand side is non-negative – its minimal value is zero. The first term, being full derivative, depends only on the boundary conditions on ϕ at $z = \pm\infty$.

Equation (3.1.24) implies that $\mathcal{E} \geq 2\,\text{Re}\left(e^{-i\eta} \Delta \mathcal{W} \right)$. The Bogomol'nyi completion can be performed with any η. However, the strongest bound is achieved provided $e^{-i\eta} \Delta \mathcal{W}$ is real. This explains the emergence of the phase factor in the BPS equations. In the model at hand, to make $e^{-i\eta} \Delta \mathcal{W}$ real, we have to choose η according to Eq. (3.1.21).

When the energy functional is written in the form (3.1.24), it is perfectly obvious that the absolute minimum is achieved provided the BPS equation (3.1.20) is satisfied. In fact, the Bogomol'nyi completion provides us with an alternative way of derivation of the BPS equations. Then, for the minimum of the energy functional – the wall tension T_w – we get

$$T_w = |\mathcal{Z}|. \qquad (3.1.25)$$

Here $\mathcal{Z}$ is the topological charge defined as

$$\mathcal{Z} = 2\left\{ \mathcal{W}(\phi(z = \infty)) - \mathcal{W}(\phi(z = -\infty)) \right\} = \frac{8\,m^3}{3\,\lambda^2}. \qquad (3.1.26)$$

In the problem at hand, the central extension of the superalgebra is tensorial, with the Lorentz structure $(1,0)+(0,1)$,

$$\left\{ Q_\alpha, Q_\beta \right\} = -4\,\Sigma_{\alpha\beta}\,\bar{\mathcal{Z}}, \qquad \left\{ \bar{Q}_{\dot{\alpha}}, \bar{Q}_{\dot{\beta}} \right\} = -4\,\bar{\Sigma}_{\dot{\alpha}\dot{\beta}}\,\mathcal{Z}. \qquad (3.1.27)$$

Here

$$\Sigma_{\alpha\beta} = -\frac{1}{2} \int dx_{[\mu} dx_{\nu]} \, (\sigma^\mu)_{\alpha\dot{\alpha}} (\bar{\sigma}^\nu)^{\dot{\alpha}}_\beta \qquad (3.1.28)$$

is the wall area tensor.

The expressions for two supercharges $\tilde{Q}_\alpha$ that do annihilate the wall are

$$\tilde{Q}_\alpha = e^{i\eta/2}\, Q_\alpha - \frac{2}{A}\, e^{-i\eta/2}\, \Sigma_{\alpha\beta}\, n^\beta_{\dot{\alpha}}\, \bar{Q}^{\dot{\alpha}}, \qquad (3.1.29)$$

where

$$n_{\alpha\dot{\alpha}} = \frac{P_{\alpha\dot{\alpha}}}{T_w A} \qquad (3.1.30)$$

is the unit vector proportional to the wall four-momentum $P_{\alpha\dot{\alpha}}$; it has only the time component in the rest frame. The subalgebra of these "residual" (unbroken) supercharges in the rest frame is

$$\left\{\tilde{Q}_\alpha, \tilde{Q}_\beta\right\} = 8 \sum_{\alpha\beta} \{T_w - |\mathcal{Z}|\}. \qquad (3.1.31)$$

The existence of the subalgebra (3.1.31) immediately proves that the wall tension T_w is equal to the central charge $\mathcal{Z}$. Indeed, $\tilde{Q}|\text{wall}\rangle = 0$ implies that $T_w - |\mathcal{Z}| = 0$. This equality is valid both to any order in perturbation theory and nonperturbatively.

From the non-renormalization theorem for the superpotential [65, 66] we additionally infer that the central charge $\mathcal{Z}$ is not renormalized. This is in contradistinction with the situation in the two-dimensional model of Section 3.1.1. The fact that in four dimensions there are more conserved supercharges than in two turns out crucial. As a consequence, the result

$$T_w = \frac{8}{3}\left|\frac{m^3}{\lambda^2}\right| \qquad (3.1.32)$$

for the wall tension is exact [62].

The wall tension T_w is a physical parameter and, as such, should be expressible in terms of the physical (renormalized) parameters m_{ren} and λ_{ren}. One can easily verify that this is compatible with the statement of non-renormalization of T_w. Indeed,

$$m = Z\, m_{\text{ren}}, \quad \lambda = Z^{3/2}\lambda_{\text{ren}},$$

where Z is the Z factor coming from the kinetic term. Consequently,

$$\frac{m^3}{\lambda^2} = \frac{m_{\text{ren}}^3}{\lambda_{\text{ren}}^2}.$$

Thus, the absence of the quantum corrections to Eq. (3.1.32), the renormalizability of the theory, and the non-renormalization theorem for superpotentials – all these three elements are intertwined with each other. In fact, every two elements taken separately imply the third one.

What lessons have we drawn from the example of the domain walls? In the centrally extended SUSY algebras the exact relation $E_{vac} = 0$ is replaced by the exact relation $T_w - |\mathcal{Z}| = 0$. Although this statement is valid both perturbatively and nonperturbatively, it is very instructive to visualize it as an explicit cancellation between bosonic and fermionic modes in perturbation theory. The non-renormalization of $\mathcal{Z}$ is a specific feature of four dimensions. We have seen previously that it does not take place in minimally supersymmetric models in two dimensions.

Finding the solution to the BPS equation

In two-dimensional theory integration of the first-order BPS equation (3.1.3) was trivial. Now the BPS equation (3.1.20) presents in fact two equations – one for the real part and one for the imaginary. Nevertheless finding the solution is still trivial. This is due to the existence of an "integral of motion,"

$$\frac{\partial}{\partial z}\left(\operatorname{Im} e^{-i\eta}\mathcal{W}\right) = 0. \tag{3.1.33}$$

The proof is straightforward and is valid in the generic Wess–Zumino model with arbitrary number of fields. Indeed, differentiating $\mathcal{W}$ and using the BPS equations we get

$$\frac{\partial}{\partial z}\left(e^{-i\eta}\mathcal{W}\right) = \left|\frac{\partial\mathcal{W}}{\partial\phi}\right|^2, \tag{3.1.34}$$

which immediately entails Eq. (3.1.33). The constraint

$$\operatorname{Im} e^{-i\eta}\mathcal{W} = \text{const} \tag{3.1.35}$$

can be interpreted as follows: in the complex $\mathcal{W}$ plane the domain wall trajectory is a straight line (see Section 3.1.1).

Living on a wall

What is the fate of two broken supercharges? As we already know, two out of four supercharges annihilate the wall – these supersymmetries are preserved in the given wall background. Two other supercharges are broken: being applied to the wall solution they create two fermion zero modes. These zero modes correspond to (2+1)-dimensional Majorana (massless) spinor field $\psi(t, x, y)$ localized on the wall.

To elucidate the above assertion it is convenient to turn first to the fate of another symmetry of the original theory which is spontaneously broken for each given wall, namely, translational invariance in the z direction.

Indeed, each wall solution, e.g. Eq. (3.1.18), breaks this invariance. This means that in fact we must deal with a family of solutions: if $\phi(z)$ is a solution, so is $\phi(z - z_0)$. The parameter z_0 is a collective coordinate – the wall center. People also refer to it as a modulus (in plural, moduli). For the static wall z_0 is a fixed constant.

Assume, however, that the wall is slightly bent. The bending should be negligible compared to the wall thickness (which is of the order of m^{-1}). The bending can be described as an adiabatically slow dependence of the wall center z_0 on t, x, and y. We will write this slightly bent wall field configuration as

$$\phi(t, x, y, z) = \phi_{\rm w}(z - \zeta(t, x, y)). \tag{3.1.36}$$

Substituting this field in the original action we arrive at the following effective (2+1)-dimensional action for the field $\zeta(t, x, y)$:

$$S_{2+1}^{\zeta} = \frac{T_{\rm w}}{2} \int d^3x \, (\partial^m \zeta)(\partial_m \zeta), \quad m = 0, 1, 2. \tag{3.1.37}$$

It is clear that $\zeta(t, x, y)$ can be viewed as a massless scalar field (called the translational modulus) which lives on the wall. It is nothing but a Goldstone field corresponding to the spontaneous breaking of the translational invariance.

Returning to two broken supercharges, they generate a Majorana (2+1)-dimensional Goldstino field $\psi_\alpha(t, x, y)$, $(\alpha = 1, 2)$ localized on the wall. The total (2+1)-dimensional effective action on the wall world volume takes the form

$$S_{2+1} = \frac{T_{\rm w}}{2} \int d^3x \left\{ (\partial^m \zeta)(\partial_m \zeta) + \bar{\psi} \partial_m \gamma^m \psi \right\} \tag{3.1.38}$$

where γ^m are three-dimensional gamma matrices (in the Majorana representation, see Appendix A, Section A.1).

The effective theory of the moduli fields on the wall worldvolume is supersymmetric, with two conserved supercharges. This is the minimal supersymmetry in 2+1 dimensions. It corresponds to the fact that two out of four supercharges are conserved.

3.1.3 D-branes in gauge field theory

In 1996 Dvali and Shifman found in supersymmetric gluodynamics [11] an anomalous $(1, 0)$ central charge in superalgebra, not seen at the classical level. They argued that this central charge is saturated by domain walls interpolating between vacua with distinct values of the order parameter, the gluino condensate $\langle \lambda\lambda \rangle$, labeling N distinct vacua of super-Yang–Mills theory with the gauge group SU(N).

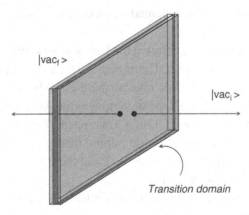

Figure 3.1. A field configuration interpolating between two distinct degenerate vacua.

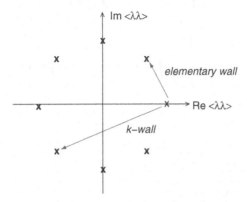

Figure 3.2. N vacua for SU(N). The vacua are labeled by the vacuum expectation value $\langle\lambda\lambda\rangle = -6\,N\,\Lambda^3\,\exp(2\pi\,i\,k/N)$ where $k = 0, 1, \ldots, N-1$. Elementary walls interpolate between two neighboring vacua.

Supersymmetric gluodynamics (it is often referred to as pure super-Yang–Mills theory) is defined by the Lagrangian

$$\mathcal{L} = \frac{1}{g^2}\int d^2\theta\,\mathrm{Tr}\,W^2 + \mathrm{H.c.} = \frac{1}{g^2}\left\{-\frac{1}{4}F^a_{\mu\nu}\,F^{a\mu\nu} + i\lambda^{a\alpha}\mathcal{D}_{\alpha\dot\beta}\bar\lambda^{a\dot\beta}\right\}, \quad (3.1.39)$$

where $\lambda^{a\alpha}$ is the Weyl spinor in the adjoint representation of SU(N).

The domain wall is a field configuration interpolating between two distinct degenerate vacua (see Fig. 3.1). There is a large variety of walls in supersymmetric gluodynamics. Minimal, or elementary, walls interpolate between vacua n and $n+1$, while k-walls interpolate between n and $n + k$, see Fig. 3.2. In [11] a mechanism was suggested for localizing gauge fields on the wall through bulk confinement.

Later this mechanism was implemented in models at weak coupling, as we will see below.

Shortly afterwards, Witten interpreted the BPS walls in supersymmetric gluody-namics as analogs of D-branes [12]. This is because their tension scales as $N \sim 1/g_s$ rather than $1/g_s^2$ typical of solitonic objects (here g_s is the string constant). Many promising consequences ensued. One of them was the Acharya–Vafa derivation of the wall worldvolume theory [67]. Using a wrapped D-brane picture and certain dualities they identified the k-wall worldvolume theory as 1+2 dimensional U(k) gauge theory with the field content of $\mathcal{N} = 2$ and the Chern–Simons term at level N breaking $\mathcal{N} = 2$ down to $\mathcal{N} = 1$.

In $\mathcal{N} = 1$ gauge theories with arbitrary matter content and superpotential the general relation (2.2.5) takes the form

$$\{Q_\alpha, Q_\beta\} = -4 \, \Sigma_{\alpha\beta} \, \bar{Z}, \tag{3.1.40}$$

where

$$\Sigma_{\alpha\beta} = -\frac{1}{2} \int dx_{[\mu} dx_{\nu]} \, (\sigma^\mu)_{\alpha\dot\alpha} (\bar\sigma^\nu)^{\dot\alpha}_\beta \tag{3.1.41}$$

is the wall area tensor, and [62, 68]

$$Z = \frac{2}{3} \Delta \left\{ \left[3\mathcal{W} - \sum_f Q_f \frac{\partial \mathcal{W}}{\partial Q_f} \right] \right.$$
$$\left. - \left[\frac{3N - \sum_f T(R_f)}{16\pi^2} \, \text{Tr} \, W^2 + \frac{1}{8} \sum_f \gamma_f \bar{D}^2 (\bar{Q}_f \, e^V Q_f) \right] \right\}_{\theta=0} \tag{3.1.42}$$

In this expression Δ implies taking the difference at two spatial infinities in the direction perpendicular to the surface of the wall. The first term in the second line presents the gauge anomaly in the central charge. The second term in the second line is a total superderivative. Therefore, it vanishes after averaging over any supersymmetric vacuum state. Hence, it can be safely omitted. The first line presents the classical result, cf. Eq. (3.1.9). At the classical level $Q_f(\partial \mathcal{W}/\partial Q_f)$ is a total superderivative too which can be seen from the Konishi anomaly [69],

$$\bar{D}^2 (\bar{Q}_f e^V Q_f) = 4 Q_f \frac{\partial \mathcal{W}}{\partial Q_f} + \frac{T(R_f)}{2\pi^2} \, \text{Tr} \, W^2. \tag{3.1.43}$$

If we discard this total superderivative for a short while (forgetting about quantum effects), we return to $Z = 2\Delta(\mathcal{W})$, the formula obtained in the Wess–Zumino model. At the quantum level $Q_f(\partial W/\partial Q_f)$ ceases to be a total superderivative because of the Konishi anomaly. It is still convenient to eliminate $Q_f(\partial W/\partial Q_f)$ in favor of $\mathrm{Tr}\,W^2$ by virtue of the Konishi relation (3.1.43). In this way one arrives at

$$Z = 2\Delta \left\{ W - \frac{N - \sum_f T(R_f)}{16\pi^2} \mathrm{Tr}\,W^2 \right\}_{\theta=0}. \qquad (3.1.44)$$

We see that the superpotential W is amended by the anomaly; in the operator form

$$W \longrightarrow W - \frac{N - \sum_f T(R_f)}{16\pi^2} \mathrm{Tr}\,W^2. \qquad (3.1.45)$$

Of course, in pure Yang–Mills theory only the anomaly term survives.

Beginning from 2002 we developed a benchmark $\mathcal{N} = 2$ model, weakly coupled in the bulk (and, thus, fully controllable), which supports both BPS walls and BPS flux tubes. We demonstrated that a gauge field is indeed localized on the wall; for the minimal wall this is a U(1) field while for non-minimal walls the localized gauge field is non-Abelian. We also found a BPS wall-string junction related to the gauge field localization, see Chapter 8. The field-theory string does end on the BPS wall, after all! The end-point of the string on the wall, after Polyakov's dualization, becomes a source of the electric field localized on the wall. In 2005 Norisuke Sakai and David Tong analyzed generic wall-string configurations. Following condensed matter physicists they called them boojums.[1]

Equation (3.1.42) implies that in pure gluodynamics (super-Yang–Mills theory without matter) the domain wall tension is

$$T = \frac{N}{8\pi^2} \left| \langle \mathrm{Tr}\lambda^2 \rangle_{\mathrm{vac\ f}} - \langle \mathrm{Tr}\lambda^2 \rangle_{\mathrm{vac\ i}} \right| \qquad (3.1.46)$$

where $\mathrm{vac}_{i,f}$ stands for the initial (final) vacuum between which the given wall interpolates. Furthermore, the gluino condensate $\langle \mathrm{Tr}\lambda^2 \rangle_{\mathrm{vac}}$ was calculated – exactly – long ago [70], using the very same methods which were later advanced and perfected by Seiberg and Seiberg and Witten in their quest for dualities in $\mathcal{N} = 1$ super-Yang–Mills theories [71] and the dual Meissner effect in $\mathcal{N} = 2$ (see [2, 3]). Namely,

$$2\langle \mathrm{Tr}\lambda^2 \rangle = \langle \lambda_\alpha^a \lambda^{a,\alpha} \rangle = -6N\Lambda^3 \exp\left(\frac{2\pi i k}{N}\right), \quad k = 0, 1, \ldots, N - 1. \quad (3.1.47)$$

---

[1] "Boojum" comes from L. Carroll's children's book *The Hunting of the Snark*. Apparently, it is fun to hunt a snark, but if the snark turns out to be a boojum, you are in trouble! Condensed matter physicists adopted the name to describe solitonic objects of the wall-string junction type in helium-3. Also: The boojum tree (Mexico) is the strangest plant imaginable. For most of the year it is leafless and looks like a giant upturned turnip. G. Sykes, found it in 1922 and said, referring to Carroll, "It must be a boojum!" The common Spanish name for this tree is Cirio, referring to its candle-like appearance.

Here k labels the N distinct vacua of the theory, see Fig. 3.2, and Λ is a dynamical scale, defined in the standard manner (i.e. in accordance with Ref. [72]) in terms of the ultraviolet parameters, M_{uv} (the ultraviolet regulator mass), and g_0^2 (the bare coupling constant),

$$\Lambda^3 = \frac{2}{3} M_{uv}^3 \left(\frac{8\pi^2}{N g_0^2}\right) \exp\left(-\frac{8\pi^2}{N g_0^2}\right). \tag{3.1.48}$$

In each given vacuum the gluino condensate scales with the number of colors as N. However, the difference of the values of the gluino condensates in two vacua which lie not too far away from each other scales as N^0. Taking into account Eq. (3.1.46) we conclude that the wall tension in supersymmetric gluodynamics

$$T \sim N.$$

(This statement just rephrases Witten's argument why the above walls should be considered as analogs of D-branes.)

The volume energy density in both vacua, to the left and to the right of the wall, vanish due to supersymmetry. Inside the transition domain, where the order parameter changes its value gradually, the volume energy density is expected to be proportional to N^2, just because there are N^2 excited degrees of freedom. Therefore, $T \sim N$ implies that the wall thickness in supersymmetric gluodynamics must scale as N^{-1}. This is very unusual, because normally we would say: the glueball mass is $O(N^0)$, hence, everything built of regular glueballs should have thickness of order $O(N^0)$.

If the wall thickness is indeed $O(N^{-1})$ the question "what consequences ensue?" immediately comes to one's mind. This issue is far from complete understanding, for relevant discussions see [73, 74, 75].

As was mentioned, there is a large variety of walls in supersymmetric gluodynamics as they can interpolate between vacua with arbitrary values of k. Even if $k_f = k_i + 1$, i.e. the wall is elementary, in fact we deal with several walls, all having one and the same tension – let us call them degenerate walls. The first indication on the wall degeneracy was obtained in Ref. [76], where two degenerate walls were observed in SU(2) theory. Later, Acharya and Vafa calculated the k-wall multiplicity [67] within the framework of D-brane/string formalism,

$$\nu_k = C_N^k = \frac{N!}{k!(N-k)!}. \tag{3.1.49}$$

For $N = 2$ only elementary walls exist, and $\nu = 2$. In the field-theoretic setting Eq. (3.1.49) was derived in [77]. The derivation is based on the fact that the index ν is topologically stable – continuous deformations of the theory do not change ν.

Thus, one can add an appropriate set of matter fields sufficient for complete Higgsing of supersymmetric gluodynamics. The domain wall multiplicity in the effective low-energy theory obtained in this way is the same as in supersymmetric gluodynamics albeit the effective low-energy theory, a Wess–Zumino type model, is much simpler.

3.1.4 Domain wall junctions

Two degenerate domain walls can coexist in one plane – a new phenomenon which, to the best of our knowledge, was first discussed in [78]. It is illustrated in Fig. 3.3. Two distinct degenerate domain walls lie on the plane; the transition domain between wall 1 and wall 2 is the domain wall junction (domain line).

Each individual domain wall is 1/2 BPS-saturated. The wall configuration with the junction line (Fig. 3.3) is 1/4 BPS-saturated. We start from $\mathcal{N} = 1$ four-dimensional bulk theory (four supercharges). Naively, the effective theory on the plane must preserve two supercharges, while the domain line must preserve one supercharge. In fact, they have four and two conserved supercharges, respectively. This is another new phenomenon – supersymmetry enhancement – discovered in [78]. One can excite the junction line endowing it with momentum in the direction of the line, without altering its BPS status. A domain line with a plane wave propagating on it (Fig. 3.3) preserves the property of the BPS saturation, see [78].

Let us pass now to more conventional wall junctions. Assume that the theory under consideration has a spontaneously broken Z_N symmetry, with $N \geq 3$, and, correspondingly, N vacua. Then one can have N distinct walls connected in the asterisk-like pattern, see Fig. 3.4. This field configuration possesses an obvious axial symmetry: the vacua are located cyclically.

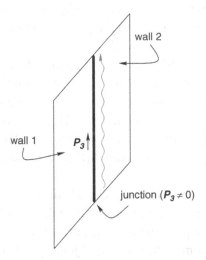

Figure 3.3. Two distinct degenerate domain walls separated by the wall junction.

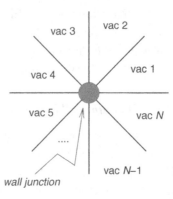

Figure 3.4. The cross section of the wall junction.

This configuration is absolutely topologically stable, as stable as the wall itself. Moreover, it can be 1/4 BPS-saturated for any value of N. It was noted [24] that theories with either U(1) or Z_N global symmetries may contain 1/4-BPS objects with axial geometry. They saturate two central charges simultaneously, (1,0) + (0,1) (the walls) and (1/2, 1/2) (the junction line).

The corresponding Bogomol'nyi equations were derived in [62] and shortly after rediscovered in [79]. Further advances in the issue of the domain wall junctions of the hub-and-spokes type were presented in [80, 81, 82, 83], see also later works [84, 85, 86, 87, 88]. We would like to single out Ref. [81] where the first analytic solution for a BPS wall junction was found in a specific generalized Wess–Zumino model. Among stimulating findings in this work is the fact that the junction tension turned out to be negative in this model. The model has Z_3 symmetry. It is derived from a SU(2) Yang–Mills theory with extended supersymmetry ($\mathcal{N} = 2$) and one matter flavor perturbed by an adjoint scalar mass. The original model contains three pairs of chiral superfields and, in addition, one extra chiral superfield. In fact, the model of [81] can be simplified and adjusted to cover the case of arbitrary N, which was done in [83]. The latter work demonstrates that the tension of the wall junctions is generically negative although exceptional models with the positive tension are possible too. Note that the negative sign of the wall junction tension does not lead to instability since the wall junctions do not exist in isolation. They are always attached to walls which stabilize this field configuration.

Returning to SU(N) supersymmetric gluodynamics ($N \geq 3$) one expects to get in this theory the 1/4 BPS junctions of the type depicted in Fig. 3.4. Of course, this theory is strongly coupled; therefore, the classical Bogomol'nyi equations are irrelevant. However, assuming that such wall junctions do exist, one can find their tension at large N even without solving the theory. To this end one uses [74, 83] the expression for the (1/2,1/2) central charge[2] in terms of the contour integral over the axial current [27]. At large N the latter integral is determined by two things: the absolute value of the gluino condensate and the overall change of the phase of the condensate when one makes the 2π rotation around the hub. In this way one arrives at the prediction

$$T_{\text{wall junction}} \sim N^2. \tag{3.1.50}$$

The coefficient in front of the N^2 factor is model dependent.

Can one interpret this N^2 dependence of the hub of the junction? Assume that each wall has thickness $1/N$ and there are N of them. Then it is natural to expect the radius of the intermediate domain where all walls join together to be of the order $(1/N) \times N \sim N^0$. This implies, in turn, that the area of the hub is $O(N^0)$. If the volume energy density inside the junction is N^2 (i.e. the same as inside the walls), one immediately gets Eq. (3.1.50).

3.1.5 Webs of walls

Domain walls can form a network when many junctions are connected together – webs or honeycombs, see Fig. 3.5 borrowed from Ref. [86]. 1/4 BPS solutions of such type were found (in the strong gauge coupling limit) in [86, 87] in four-dimensional $\mathcal{N} = 2$ supersymmetric Yang–Mills theory with the gauge

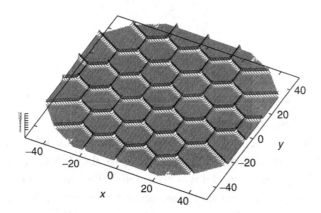

Figure 3.5. Honeycomb web of domain walls. This web in this figure divides 37 vacua and has 18 external legs and 19 internal faces. The moduli space corresponds to CP(36) whose dimension is 72.

---

[2] There is a subtle point here which must be noted. For the wall type of the hub-and-spokes type the overall tension is the sum of two tensions: the tension of the walls and the tension of the hub. The first is determined by the (1,0) central charge, the second by (1/2,1/2). Each separately is somewhat ambiguous in the case at hand. The ambiguity cancels in the sum [27].

group U(N_c) and N_f flavor hypermultiplets in the fundamental representation ($N_f > N_c$). This model is described in detail in Sections 4.1 and 4.7. The solution saturates two central charges, (1,0) + (0,1) and (1/2,1/2). The moduli space of this particular web of walls is the complex Grassmann manifold $G_{N_f,N_c} = \text{SU}(N_f)/[\text{SU}(N_f - N_c) \times \text{SU}(N_c) \times \text{U}(1)]$.

The web of walls can contain several external legs and loops whose maximal numbers are determined by N_f and N_c. If the gauge group is U(1) rather than U(N_c) (with $N_c \geq 2$) the moduli space of the web of walls simplifies and becomes CP($N_f - 1$).

Further studies of dynamics of the domain wall loops, as in Fig. 3.5, were carried out in [89]. The authors used the moduli approximation and found that a phase rotation induces a repulsive force which can be interpreted as a Noether charge of Q solitons.

3.2 Vortices in $D = 3$ and flux tubes in $D = 4$

Vortices were among the first examples of topological defects treated in the Bogomol'nyi limit [5, 4, 1] (see also [90]). Explicit embedding of the bosonic sector in supersymmetric models dates back to the 1980s. In [91] a three-dimensional Abelian Higgs model was considered. That model had $\mathcal{N} = 1$ supersymmetry (two supercharges) and thus, according to Section 2.2.2, contained no central charge that could be saturated by vortices. Hence, the vortices discussed in [91] were noncritical. BPS saturated vortices can and do occur in $\mathcal{N} = 2$ three-dimensional models (four supercharges) with a non-vanishing Fayet–Iliopoulos term [92, 93]. Such models can be obtained by dimensional reduction from four-dimensional $\mathcal{N} = 1$ models. We will start from a brief excursion in SQED.

3.2.1 SQED in 3D

The starting point is SQED with the Fayet–Iliopoulos term ξ in four dimensions. The SQED Lagrangian is

$$\mathcal{L} = \left\{ \frac{1}{4 e^2} \int d^2\theta \, W^2 + \text{H.c.} \right\} + \int d^4\theta \, \bar{Q} \, e^{n_e V} \, Q$$

$$+ \int d^4\theta \, \bar{\tilde{Q}} \, e^{-n_e V} \, \tilde{Q} - n_e \xi \int d^2\theta d^2\bar{\theta} \, V(x, \theta, \bar{\theta}), \qquad (3.2.1)$$

where e is the electric coupling constant, Q and $\tilde{Q}$ are chiral matter superfields (with charges n_e and $-n_e$, respectively), and W_α is the supergeneralization of the photon field strength tensor,

$$W_\alpha = \frac{1}{8} \bar{D}^2 \, D_\alpha V = i \left(\lambda_\alpha + i\theta_\alpha D - \theta^\beta F_{\alpha\beta} - i\theta^2 \partial_{\alpha\dot\alpha} \bar{\lambda}^{\dot\alpha} \right). \qquad (3.2.2)$$

In four dimensions the absence of the chiral anomaly in SQED requires the matter superfields enter in pairs of the opposite charges, e.g.

$$i\mathcal{D}_\mu \psi = \left(i\partial_\mu + n_e A_\mu \right) \psi, \quad i\mathcal{D}_\mu \tilde{\psi} = \left(i\partial_\mu - n_e A_\mu \right) \tilde{\psi}. \qquad (3.2.3)$$

Otherwise the theory is anomalous, the chiral anomaly renders it non-invariant under gauge transformations. Thus, the minimal matter sector includes two chiral superfields Q and $\tilde{Q}$, with charges n_e and $-n_e$, respectively. (In the literature a popular choice is $n_e = 1$. In Part II we will use a different normalization, $n_e = 1/2$, which is more convenient in some problems that we address in Part II.)

In three dimensions there is no chirality. Therefore, one can consider 3D SQED with a single matter superfield Q, with charge n_e. Classically it is perfectly fine just to discard the superfield $\tilde{Q}$ from the Lagrangian (3.2.1). However, such "crudely truncated" theory may be inconsistent at the quantum level [94, 95, 96]. Gauge invariance in loops requires, as we will see shortly, simultaneous introduction of a Chern–Simons term in the one matter superfield model [94, 95, 96]. The Chern–Simons term breaks parity. That's the reason why this phenomenon is sometimes referred to as parity anomaly.

A perfectly safe way to get rid of $\tilde{Q}$ is as follows. Let us start from the two-superfield model (3.2.1), which is certainly self-consistent both at the classical and quantum levels. The one-superfield model can be obtained from that with two superfields by making $\tilde{Q}$ heavy and integrating it out. If one manages to introduce a mass $\tilde{m}$ for $\tilde{Q}$ without breaking $\mathcal{N} = 2$ supersymmetry, the large $\tilde{m}$ limit can be viewed as an excellent regularization procedure.

Such mass terms are well known, for a review see [97, 98, 96]. They go under the name of "real masses," are specific to theories with U(1) symmetries dimensionally reduced from $D = 4$ to $D = 3$, and present a direct generalization of twisted masses in two dimensions [32]. To introduce a "real mass" one couples matter fields to a background vector field with a non-vanishing component along the reduced direction. For instance, in the case at hand we introduce a background field V_b as

$$\Delta \mathcal{L}_m = \int d^4\theta \, \bar{\tilde{Q}} \, e^{V_b} \, \tilde{Q}, \quad V_b = \tilde{m} \, (2\,i) \left(\theta^1 \bar{\theta}^{\dot{2}} - \theta^2 \bar{\theta}^{\dot{1}} \right). \quad (3.2.4)$$

The reduced spatial direction is that along the y axis. We couple V_b to the U(1) current of $\tilde{Q}$ ascribing to $\tilde{Q}$ charge one with respect to the background field. At the same time Q is assumed to have V_b charge zero and, thus, has no coupling to V_b. Then, the background field generates a mass term only for $\tilde{Q}$, without breaking $\mathcal{N} = 2$. Needless to say, there is no kinetic term for V_b. Equation (3.2.4) implies that $\tilde{m} = (A_b)_2$.

After reduction to three dimensions and passing to components (in the Wess–Zumino gauge) we arrive at the action in the following form (in the three-dimensional notation):

$$S = \int d^3x \left\{ -\frac{1}{4e^2} F_{\mu\nu} F^{\mu\nu} + \frac{1}{2e^2} (\partial_\mu a)^2 + \frac{1}{e^2} \bar{\lambda} \, i \slashed{\partial} \lambda \right.$$
$$+ \frac{1}{2e^2} D^2 - n_e \xi D + n_e D \left(\bar{q} \, q - \bar{\tilde{q}} \, \tilde{q} \right)$$
$$+ \left[D^\mu \bar{q} \, D_\mu q + \bar{\psi} \, i \slashed{D} \psi \right] + \left[D^\mu \bar{\tilde{q}} \, D_\mu \tilde{q} + \bar{\tilde{\psi}} \, i \slashed{D} \tilde{\psi} \right]$$
$$- a^2 \bar{q} \, q - (\tilde{m} + a)^2 \bar{\tilde{q}} \, \tilde{q} + a \, \bar{\psi} \, \psi - (\tilde{m} + a) \, \bar{\tilde{\psi}} \, \tilde{\psi}$$
$$+ n_e \left[\sqrt{2} \left(\bar{\lambda} \, \psi \right) \bar{q} + \text{H.c.} \right] - n_e \left[\sqrt{2} \left(\bar{\lambda} \, \tilde{\psi} \right) \bar{\tilde{q}} + \text{H.c.} \right] \right\}. \quad (3.2.5)$$

Here a is a real scalar field,

$$a = -n_e A_2,$$

λ is the photino field, and q, $\tilde{q}$ and ψ, $\tilde{\psi}$ are matter fields belonging to Q and $\tilde{Q}$, respectively. The covariant derivatives are defined in Eq. (3.2.3). Finally, D is an auxiliary field, the last component of the superfield V. Eliminating D via the equation of motion we get the scalar potential

$$V = \frac{e^2}{2} n_e^2 \left[\xi - \left(\bar{q} \, q - \bar{\tilde{q}} \, \tilde{q} \right) \right]^2 + a^2 \bar{q} \, q + (\tilde{m} + a)^2 \bar{\tilde{q}} \, \tilde{q}, \quad (3.2.6)$$

which implies a potentially rather rich vacuum structure. For our purposes – the BPS-saturated vortices – only the Higgs phase is of importance. We will assume that

$$\xi > 0, \quad \tilde{m} \geq 0. \quad (3.2.7)$$

If $\tilde{\psi}$ and $\tilde{q}$ are viewed as regulators (i.e. $\tilde{m} \to \infty$), they can be integrated out leaving us with the one matter superfield model. It is obvious that integrating them out we get a Chern–Simons term at one loop,[3] with a well-defined coefficient that does not vanish in the limit $\tilde{m} = \infty$. We prefer to keep $\tilde{m}$ as a free parameter, assuming that $\tilde{m} \neq 0$.

From the standpoint of vortex studies, the model (3.2.1) *per se* is not quite satisfactory due to the existence of the flat direction (correspondingly, there is a gapless mode which renders the theory ill-defined in the infrared, see Section 5.1). The flat direction is eliminated at $\tilde{m} \neq 0$. Thus, there are three relevant parameters of dimension of mass,

$$e^2, \; \xi, \; \text{and} \; \tilde{m}.$$

The weak coupling regime implies that $e^2/\xi \ll 1$.

If $\tilde{m} \neq 0$ the vacuum field configuration is as follows:

$$\tilde{q} = 0, \quad a = 0, \quad \bar{q}q = \xi. \tag{3.2.8}$$

The vanishing of the D term in the vacuum requires $\bar{q} \, q_{\text{vac}} = \xi$. Then the term $a^2 \bar{q}q$ in (3.2.6) implies that $a = 0$ in the vacuum. Up to gauge transformations the vacuum is unique. The Higgs phase is enforced by our choice $\tilde{m} \neq 0$ and $\xi \neq 0$. The fields $\tilde{q}$, $\tilde{\psi}$ play a role only at the level of quantum corrections, providing a well-defined regularization in loops.

Central charge

The general form of the centrally extended $\mathcal{N} = 2$ superalgebra in $D = 3$ was discussed in Section 2.3.2. The central charge relevant in the problem at hand – vortices – is presented by the last term in Eq. (2.3.6). It can be conveniently derived using the complex representation for supercharges and reducing from $D = 4$ to $D = 3$. In four dimensions [27]

$$\{Q_\alpha, \bar{Q}_{\dot{\alpha}}\} = 2P_{\alpha\dot{\alpha}} + 2Z_{\alpha\dot{\alpha}} \equiv 2\left(P_\mu + Z_\mu\right)\left(\sigma^\mu\right)_{\alpha\dot{\alpha}}, \tag{3.2.9}$$

where P_μ is the momentum operator, and

$$Z_\mu = n_e \, \xi \int d^3x \, \epsilon_{0\mu\nu\rho} \left(\partial^\nu A^\rho\right) + \cdots \tag{3.2.10}$$

Here ellipses denote full spatial derivatives of currents [4] that fall off exponentially fast at infinity. Such terms are clearly inessential.

---

[3] In passing from two matter superfields to one, in order to justify integrating out $\tilde{Q}$, one must consider $\tilde{m} \gg e\sqrt{\xi}$. Given that $e^2/\xi \ll 1$, the condition $\tilde{m} \gg e\sqrt{\xi}$ does not necessarily imply that $\tilde{m} \gg \xi$.
[4] Moreover, these currents are not unambiguously defined, see [27].

In three dimensions the central charge of interest reduces to $P_2 + Z_2$. Thus, in terms of complex supercharges the appropriate centrally extended algebra takes the form[5]

$$\left\{ Q, \left(Q^\dagger \right) \gamma^0 \right\} = 2 \left(P_0 \gamma^0 + P_1 \gamma^x + P_3 \gamma^z \right)$$

$$+ 2 \left\{ \frac{1}{e^2} \int d^2 x \vec{\nabla} \left(\vec{E} \, a \right) + \tilde{m} \, q - n_e \xi \int d^2 x \, B \right\}, \quad (3.2.11)$$

where $\vec{E}$ is the electric field, B is the magnetic field,

$$B = \frac{\partial A_z}{\partial x} - \frac{\partial A_x}{\partial z}, \quad (3.2.12)$$

and q is a conserved Noether charge,

$$q = \int d^2 x \, j^0, \quad j^\mu \equiv \bar{\tilde{\psi}} \gamma^\mu \tilde{\psi} + \bar{\tilde{q}} \, i \overset{\leftrightarrow}{\mathcal{D}}_\mu \tilde{q}. \quad (3.2.13)$$

The second line in Eq. (3.2.11) presents the vortex-related central charge.[6] The term proportional to a gives a vanishing contribution to the central charge. However, the q term (sometimes omitted in the literature) plays an important role. It combines with the ξ term in the expression for the vortex mass converting the bare value of ξ into the renormalized one. In the problem at hand, the vortex mass gets renormalized at one loop, and so does the Fayet–Iliopoulos parameter.

BPS equation for the vortex

At the classical level the fields a and $\tilde{q}$ play no role. They will be set

$$\tilde{q} = 0, \quad a = 0. \quad (3.2.14)$$

The first-order equations describing the ANO vortex in the Bogomol'nyi limit [5, 4, 1] take the form

$$B - n_e \, e^2 (|q|^2 - \xi) = 0,$$
$$(\mathcal{D}_x + i \mathcal{D}_z) \, q = 0, \quad (3.2.15)$$

with the boundary conditions

$$q \to \sqrt{\xi} \, e^{ik\alpha} \quad \text{at} \quad r \to \infty,$$

$$q \to 0 \quad \text{at} \quad r \to 0, \quad (3.2.16)$$

---

[5] In the following expression terms containing equations of motion of the type $a(\vec{\nabla} \vec{E} - J_0)$ are omitted.
[6] The emergence of the U(1) Noether charge $\tilde{m} q$ in the central charge is in one-to-one correspondence with a similar phenomenon taking place in the two-dimensional CP($N-1$) models with the twisted mass [34].

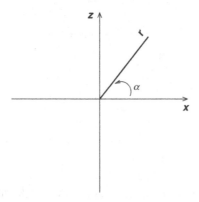

Figure 3.6. Polar coordinates on the x, z plane.

where α is the polar angle on the x, z plane, while r is the distance from the origin in the same plane (Fig. 3.6). Moreover k is an integer, counting the number of windings.

If Eqs. (3.2.15) are satisfied, the flux of the magnetic field is $2\pi k$ (the winding number k determines the quantized magnetic flux), and the vortex mass (string tension) is

$$M = 2\pi \xi k, \tag{3.2.17}$$

The linear dependence of the k-vortex mass on k implies the absence of their potential interaction.

For the elementary $k = 1$ vortex it is convenient to introduce two profile functions $\phi(r)$ and $f(r)$ as follows:

$$q(x) = \phi(r) e^{i\alpha}, \quad A_n(x) = -\frac{1}{n_e} \varepsilon_{nm} \frac{x_m}{r^2} [1 - f(r)] . \tag{3.2.18}$$

The *ansatz* (3.2.18) goes through the set of equations (3.2.15), and we get the following two equations on the profile functions:

$$-\frac{1}{r}\frac{df}{dr} + n_e^2 e^2 \left(\phi^2 - \xi\right) = 0, \quad r\frac{d\phi}{dr} - f\phi = 0. \tag{3.2.19}$$

The boundary conditions for the profile functions are rather obvious from the form of the *ansatz* (3.2.18) and from our previous discussion. At large distances

$$\phi(\infty) = \sqrt{\xi}, \quad f(\infty) = 0. \tag{3.2.20}$$

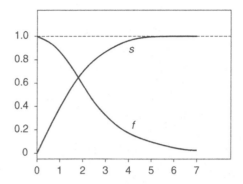

Figure 3.7. Profile functions of the string as functions of the dimensionless variable $m_\gamma r$. The gauge and scalar profile functions are given by f and $s \equiv \phi/\sqrt{\xi}$, respectively.

At the same time, at the origin the smoothness of the field configuration at hand (the absence of singularities) requires

$$\phi(0) = 0, \quad f(0) = 1. \tag{3.2.21}$$

These boundary conditions are such that the scalar field reaches its vacuum value at infinity. Equations (3.2.19) with the above boundary conditions lead to a unique solution for the profile functions, although its analytic form is not known. The vortex size is $\sim e^{-1}\xi^{-1/2}$. The solution can be readily obtained numerically. The profile functions ϕ and f which determine the Higgs field and the gauge potential, respectively, are shown in Fig. 3.7.

The fermion zero modes

Quantization of vortices requires the knowledge of the fermion zero modes for the given classical solution. More precisely, since the solution under consideration is static, we are interested in the zero-eigenvalue solutions of the static fermion equations which, thus, effectively become two- rather than three-dimensional,

$$i\left(\gamma^x \mathcal{D}_x + \gamma^z \mathcal{D}_z\right)\psi + n_e \sqrt{2}\lambda q = 0, \tag{3.2.22}$$
$$i\left(\gamma^x \partial_x + \gamma^z \partial_z\right)\lambda + e^2 n_e \sqrt{2}\psi \bar{q} = 0.$$

These equations are obtained from (3.2.5) where we dropped the tilded terms (since $\tilde{q} = 0$). The fermion operator is Hermitean implying that every solution for $\{\psi, \lambda\}$ is accompanied by that for $\{\bar{\psi}, \bar{\lambda}\}$.

Since the solution to equations (3.2.15) discussed above is 1/2 BPS, two of the four supercharges annihilate it while the other two generate the fermion zero modes – superpartners of translational modes. One can show [99] that these are the only normalizable fermion zero modes in the problem at hand.

Short versus long representations

The (1+2)-dimensional model under consideration has four supercharges. The corresponding regular super-representation is four-dimensional (i.e. contains two bosonic and two fermionic states).

The vortex we discuss has two fermion zero modes. Hence, viewed as a particle in 1+2 dimensions it forms a super-doublet (one bosonic state plus one fermionic). Hence, this is a short multiplet. This implies, of course, that the BPS bound must remain saturated when quantum corrections are switched on. Both the central charge and the vortex mass get corrections [100, 99], but they remain equal to each other.

Vortex mass and central charge renormalizations

Assuming that $n_e = 1$ and saturating the central charge in Eq. (3.2.11) by the vortex soliton we get

$$Z_{\text{vortex}} = -\xi \int d^2x\, B + \tilde{m}q = -2\pi\, \xi + \frac{\tilde{m}}{2}. \qquad (3.2.23)$$

Here we use the fact that the induced q charge of the vortex is 1/2. This is not difficult to see for any value of $\tilde{m}$ [101]. Proving this assertion becomes especially simple at large $\tilde{m}$ when one can just integrate the tilded fields out in the given vortex field. One then arrives at

$$q = \int d^2x\, \frac{1}{4\pi}\, B = \frac{1}{2}. \qquad (3.2.24)$$

Since the renormalized value of the FI parameter ξ is

$$\xi_R = \xi + \frac{m_q - \tilde{m}}{4\pi} \qquad (3.2.25)$$

where $m_q = \sqrt{2\xi}\, e$ is the mass of the untilded particles, we can rewrite Eq. (3.2.23) in the form

$$Z_{\text{vortex}} = -2\pi\left(\xi_R - \frac{m_q}{4\pi}\right). \qquad (3.2.26)$$

In the very same "physical" regularization scheme outlined above the vortex mass shifts by the same amount [100, 101], and

$$M_{\text{vortex}} = 2\pi\left(\xi_R - \frac{m_q}{4\pi}\right) = |Z_{\text{vortex}}|. \qquad (3.2.27)$$

3.2.2 *Four-dimensional SQED and the ANO string*

In this section we will discuss $\mathcal{N} = 1$ SQED. SQED with extended supersymmetry (i.e. $\mathcal{N} = 2$) is also very interesting. This latter model is presented in Appendix C.

The Lagrangian is the same as in Eq. (3.2.1). We will consider the simplest case: one chiral superfield Q with charge $n_e = 1/2$, and one chiral superfield $\tilde{Q}$ with charge $n_e = -1/2$. The electric charge of matter is chosen to be half-integer to make contact with what follows. This normalization is convenient in the case of non-Abelian models, see Part II. The Lagrangian in components can be obtained from Eq. (3.2.5) by setting $a = \tilde{m} = 0$. The scalar potential obviously takes the form

$$V = \frac{e^2}{2} n_e^2 \left[\xi - \left(\bar{q}\, q - \bar{\tilde{q}}\, \tilde{q} \right) \right]^2 . \tag{3.2.28}$$

The vacuum manifold is a "hyperboloid"

$$\bar{q}\, q - \bar{\tilde{q}}\, \tilde{q} = \xi . \tag{3.2.29}$$

Thus, we deal with the Higgs branch of real dimension two. In fact, the vacuum manifold can be parametrized by a complex modulus $\tilde{q}q$. On this Higgs branch the photon field and superpartners form a massive supermultiplet, while $\tilde{q}q$ and superpartners form a massless one.

As was shown in [102], no finite-thickness vortices exist at a generic point on the vacuum manifold, due to the absence of the mass gap (presence of the massless Higgs excitations). The moduli fields get involved in the solution at the classical level generating a logarithmically divergent tail. An infrared regularization can remove this logarithmic divergence, and vortices become well-defined, see [103] and Chapter 7. One of the possible infrared regularizations is considering a finite-length string instead of an infinite string. Then all infrared divergences are cut off at distances of the order of the string length. The thickness of the string is of the order of logarithm of this length. This is discussed in detail in Chapter 7. Needless to say, such string is not BPS-saturated.

At the base of the Higgs branch, at $\tilde{q} = 0$, the classical solutions of the BPS equations for q and A_μ are well-defined. The form of the solution coincides with that given in Section 3.2.1.

The fact that there is a flat direction and, hence, massless particles in the bulk theory does not disappear, of course. Even though at $\tilde{q} = 0$ the classical string solution is well-defined, infrared problems arise at the loop level. One can avoid massless particles in the spectrum if one embeds the theory (3.2.5) in SQED with eight supercharges, see Section 5.1 and Appendix C. Then the Higgs branch is eliminated, and one is left with isolated vacua. After the embedding is done, one can break $\mathcal{N} = 2$ down to $\mathcal{N} = 1$, if one so desires.

A simpler framework is provided by the so-called M model. Its non-Abelian version is considered in Section 5.2. Here we will outline the construction of this model in the context of $\mathcal{N} = 1$ SQED.

We introduce an extra *neutral* chiral superfield M, which interacts with Q and $\tilde{Q}$ through the super-Yukawa coupling,

$$\mathcal{L}_M = \int d^2\theta \, d^2\bar{\theta} \, \frac{1}{h} \, \bar{M} \, M + \left\{ \int d^2\theta \, QM\tilde{Q} + (\text{H.c}) \right\}. \qquad (3.2.30)$$

Here h is a coupling constant. As we will see momentarily the Higgs branch is lifted. An obvious advantage of this model is that it makes no reference to $\mathcal{N} = 2$. This is probably the simplest $\mathcal{N} = 1$ model which supports BPS-saturated ANO strings without infrared problems.

The scalar potential (3.2.28) is now replaced by

$$V_M = \frac{e^2}{2} \, n_e^2 \left[\xi - \left(\bar{q} \, q - \bar{\tilde{q}} \, \tilde{q} \right) \right]^2 + h \, |q \, \tilde{q}|^2 + |q \, M|^2 + |M \, \tilde{q}|^2 . \qquad (3.2.31)$$

The vacuum is unique modulo gauge transformations,

$$q = \bar{q} = \sqrt{\xi} \,, \quad \tilde{q} = 0 \,, \quad M = 0. \qquad (3.2.32)$$

The classical ANO flux tube solution considered above remains valid as long as we set, additionally, $\tilde{q} = M = 0$. The string tension is the same, $T_{\text{string}} = 2\pi\xi$. (Note that in Eq. (3.2.31) the parameter ξ is defined with n_e^2 factored out. See also Eq. (C.11) and its derivation.) The quantization procedure is straightforward, since one encounters no infrared problems whatsoever – all particles in the bulk are massive. In particular, there are four normalizable fermion zero modes (cf. Ref. [35]).

For further thorough discussions we refer the reader to Section 7.2.

3.2.3 Flux tube junctions

In theories with Z_N symmetry the ANO flux tubes can form junctions of the type depicted in Fig. 3.8. As an example, let us consider a U(1) × U(1) × U(1) gauge theory with three "photons" and three (scalar) matter fields, ϕ, χ, and η,

$$\mathcal{L} = -\frac{1}{4e^2} \sum_{i=1}^{3} (F_i)_{\mu\nu} (F_i)^{\mu\nu} + (D_\mu\bar{\phi})(D^\mu\phi) + (D_\mu\bar{\chi})(D^\mu\chi)$$
$$+ (D_\mu\bar{\eta})(D^\mu\eta) + V(\phi, \chi, \eta), \qquad (3.2.33)$$

Table 3.1. *Couplings of ϕ, χ, and η with respect to three photons A_1, A_2, and A_3 of the $U(1)^3$ theory (3.2.33).*

	ϕ	χ	η
A_1	2/3	2/3	−1/3
A_2	2/3	−1/3	2/3
A_3	−1/3	2/3	2/3

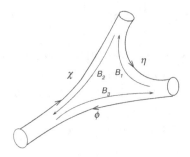

Figure 3.8. A junction of three flux tubes ("Mercedes logo") in the Z_3 invariant theory (3.2.33). The letters ϕ, χ, and η show which fields have windings in three sectors.

whose electric charges with respect to three photons are presented in Table 3.1. The potential $V(\phi, \chi, \eta)$ is assumed to be symmetric under the interchange of ϕ, χ, and η. Another requirement to $V(\phi, \chi, \eta)$ is spontaneous breaking of all three U(1) gauge groups through nonvanishing expectation values $\langle\phi\rangle = \langle\chi\rangle = \langle\eta\rangle \neq 0$.

The three flux tubes form a planar structure of the "Mercedes logo" type, with $2\pi/3$ angles between them. The flux tube in the left-hand side of Fig. 3.8 carries the magnetic fluxes of the third and second photons, the next (clockwise) flux tube the magnetic fluxes of the first and second photons, and the last flux tube of the first and third.

3.3 Monopoles

In this section we will discuss magnetic monopoles – very interesting objects which carry magnetic charges. They emerge as free magnetically charged particles in non-Abelian gauge theories in which the gauge symmetry is spontaneously broken down to an Abelian subgroup.[7] The simplest example was found by 't Hooft [105] and Polyakov [106]. The model they considered had been invented by Georgi and Glashow [107] for different purposes. As it often happens, the Georgi–Glashow model turned out to be more valuable than the original purpose, which is long forgotten, while the model itself is alive and well and is being constantly used by theorists.

3.3.1 The Georgi–Glashow model: vacuum and elementary excitations

Let us begin with a brief description of the Georgi–Glashow model. The gauge group is SU(2) and the matter sector consists of one real scalar field ϕ^a in the adjoint representation (i.e. SU(2) triplet). The Lagrangian of the model is

$$L = -\frac{1}{4g^2} F^a_{\mu\nu} F^{\mu\nu,a} + \frac{1}{2}(D_\mu \phi^a)(D^\mu \phi^a) - \frac{1}{8}\lambda(\phi^a \phi^a - v^2)^2, \quad (3.3.1)$$

where the covariant derivative in the adjoint acts as

$$D_\mu \phi^a = \partial_\mu \phi^a + \varepsilon^{abc} A^b_\mu \phi^c. \quad (3.3.2)$$

Below we will focus on the limit of BPS monopoles. This limit corresponds to a vanishing scalar coupling, $\lambda \to 0$. The only role of the last term in Eq. (3.3.1) is to provide a boundary condition for the scalar field. As is clear from Chapter 2 the monopole central charge exists only in $\mathcal{N} = 2$ and $\mathcal{N} = 4$ superalgebras. Therefore, one should understand the theory (3.3.1) (at $\lambda = 0$) as embedded in super-Yang–Mills theories with extended superalgebra. In Part II we will extensively discuss such embeddings in the context of $\mathcal{N} = 2$.

The classical definition of magnetic charges refers to theories that support a long-range (Coulomb) magnetic field. Therefore, in consideration of the isolated monopole the pattern of the symmetry breaking should be such that some of the gauge bosons remain massless. In the Georgi–Glashow model (3.3.1) the pattern is as follows:

$$\text{SU}(2) \to \text{U}(1). \quad (3.3.3)$$

---

[7] In the confining regime monopoles can be obtained in some theories with no adjoint fields, in which the gauge symmetry is broken completely [104]. This is a recent development.

To see that this is indeed the case let us note the ϕ^a self-interaction term (the last term in Eq. (3.3.1)) forces ϕ^a to develop a vacuum expectation value,

$$\langle \phi^a \rangle = v\delta^{3a}. \tag{3.3.4}$$

The direction of the vector ϕ^a in the SU(2) space (to be referred to as "color space" or "isospace") can be chosen arbitrarily. One can always reduce it to the form (3.3.4) by a global color rotation. Thus, Eq. (3.3.4) can be viewed as a (unitary) gauge condition on the field ϕ.

   This gauge is very convenient for discussing the particle content of the theory, elementary excitations. Since the color rotation around the third axis does not change the vacuum expectation value of ϕ^a,

$$\exp\left\{ i\alpha \frac{\tau_3}{2} \right\} \phi_{\text{vac}} \exp\left\{ -i\alpha \frac{\tau_3}{2} \right\} = \phi_{\text{vac}}, \quad \phi_{\text{vac}} = v\frac{\tau_3}{2}, \tag{3.3.5}$$

the third component of the gauge field remains massless – we will call it a "photon,"

$$A_\mu^3 \equiv A_\mu, \quad F_{\mu\nu} = \partial_\mu A_\nu - \partial_\nu A_\mu. \tag{3.3.6}$$

The first and the second components form massive vector bosons,

$$W_\mu^\pm = \frac{1}{\sqrt{2}\,g} \left(A_\mu^1 \pm A_\mu^2 \right). \tag{3.3.7}$$

As usual in the Higgs mechanism, the massive vector bosons eat up the first and the second components of the scalar field ϕ^a. The third component, the physical Higgs field, can be parametrized as

$$\phi^3 = v + \varphi, \tag{3.3.8}$$

where φ is the physical Higgs field. In terms of these fields the Lagrangian (3.3.1) can be readily rewritten as

$$\begin{aligned}
L = &-\frac{1}{4g^2} F_{\mu\nu} F_{\mu\nu} + \frac{1}{2}(\partial_\mu \varphi)^2 \\
&- \left(D_\alpha W_\mu^+ \right) \left(D_\alpha W_\mu^- \right) + \left(D_\mu W_\mu^+ \right) \left(D_\nu W_\nu^- \right) + g^2 (v + \phi)^2\, W_\mu^+ W_\mu^- \\
&- 2\, W_\mu^+ F_{\mu\nu} W_\nu^- + \frac{g^2}{4} \left(W_\mu^+ W_\nu^- - W_\nu^+ W_\mu^- \right)^2,
\end{aligned} \tag{3.3.9}$$

where the covariant derivative now includes only the photon field,

$$D_\alpha W^\pm = \left(\partial_\alpha \pm i A_\alpha \right) W^\pm. \tag{3.3.10}$$

The last line presents the magnetic moment of the charged (massive) vector bosons and their self-interaction. In the limit $\lambda \to 0$ the physical Higgs field is massless. The mass of the $W^{\pm}$ bosons is

$$M_W = g\,v. \tag{3.3.11}$$

3.3.2 Monopoles – topological argument

Let us explain why this model has a topologically stable soliton.

Assume that the monopole's center is at the origin and consider a large sphere $\mathcal{S}_R$ of radius R with the center at the origin. Since the mass of the monopole is finite, by definition, $\phi^a \phi^a = v^2$ on this sphere. ϕ^a is a three-component vector in the isospace subject to the constraint $\phi^a \phi^a = v^2$ which gives us a two-dimensional sphere $\mathcal{S}_G$. Thus, we deal here with mappings of $\mathcal{S}_R$ into $\mathcal{S}_G$. Such mappings split in distinct classes labeled by an integer n, counting how many times the sphere $\mathcal{S}_G$ is swept when we sweep once the sphere $\mathcal{S}_R$, since

$$\pi_2(\mathrm{SU}(2)/\mathrm{U}(1)) = Z. \tag{3.3.12}$$

$\mathcal{S}_G = \mathrm{SU}(2)/\mathrm{U}(1)$ because for each given vector ϕ^a there is a U(1) subgroup which does *not* rotate it. The SU(2) group space is a three-dimensional sphere while that of SU(2)/U(1) is a two-dimensional sphere.

An isolated monopole field configuration (the 't Hooft–Polyakov monopole) corresponds to a mapping with $n = 1$. Since it is impossible to continuously deform it to the topologically trivial mapping, the monopoles are topologically stable.

3.3.3 Mass and magnetic charge

Classically the monopole mass is given by the energy functional

$$E = \int d^3x \left\{ \frac{1}{2\,g^2}\, B_i^a B_i^a + \frac{1}{2}\left(D_i \phi^a\right)\left(D_i \phi^a\right) \right\}, \tag{3.3.13}$$

where

$$B_i^a = -\frac{1}{2}\, \varepsilon_{ijk} F_{jk}^a. \tag{3.3.14}$$

The fields are assumed to be time-independent, $B_i^a = B_i^a(\vec{x})$, $\phi^a = \phi^a(\vec{x})$. For static fields it is natural to assume that $A_0^a = 0$. This assumption will be verified a posteriori, after we find the field configuration minimizing the functional (3.3.13). Equation (3.3.13) assumes the limit $\lambda \to 0$. However, in performing minimization we should keep in mind the boundary condition $\phi^a(\vec{x})\phi^a(\vec{x}) \to v^2$ at $|\vec{x}| \to \infty$.

Equation (3.3.13) can be identically rewritten as follows:

$$E = \int d^3x \left\{ \frac{1}{2} \left(\frac{1}{g} B_i^a - D_i \phi^a \right) \left(\frac{1}{g} B_i^a - D_i \phi^a \right) + \frac{1}{g} B_i^a D_i \phi^a \right\}. \qquad (3.3.15)$$

The last term on the right-hand side is a full derivative. Indeed, after integrating by parts and using the equation of motion $D_i B_i^a = 0$ we get

$$\int d^3x \left\{ \frac{1}{g} B_i^a D_i \phi^a \right\} = \frac{1}{g} \int d^3x \, \partial_i \left(B_i^a \phi^a \right)$$

$$= \frac{1}{g} \int_{S_R} d^2 S_i \left(B_i^a \phi^a \right). \qquad (3.3.16)$$

In the last line we made use of Gauss' theorem and passed from the volume integration to that over the surface of the large sphere. Thus, the last term in Eq. (3.3.15) is topological.

The combination $B_i^a \phi^a$ can be viewed as a gauge invariant definition of the magnetic field $\vec{B}$. More exactly,

$$\mathcal{B}_i = \frac{1}{v} B_i^a \phi^a. \qquad (3.3.17)$$

Indeed, far away from the monopole core one can always assume ϕ^a to be aligned in the same way as in the vacuum (in an appropriate gauge), $\phi^a = v \delta^{3a}$. Then $\mathcal{B}_i = B_i^3$. The advantage of the definition (3.3.17) is that it is gauge independent.

Furthermore, the magnetic charge Q_M inside a sphere S_R can be defined through the flux of the magnetic field through the surface of the sphere,[8]

$$Q_M = \int_{S_R} d^2 S_i \, \frac{1}{g} \mathcal{B}_i. \qquad (3.3.18)$$

From Eq. (3.3.30) (see below) we will see that

$$\mathcal{B}_i \equiv \frac{1}{v} B_i^a \phi^a \longrightarrow n^i \frac{1}{r^2} \quad \text{at } r \to \infty, \qquad (3.3.19)$$

and, hence,

$$Q_M = \frac{4\pi}{g}. \qquad (3.3.20)$$

---

[8] A remark: Conventions for the charge normalization used in different books and papers may vary. In his original paper on the magnetic monopole [108], Dirac uses the convention $e^2 = \alpha$ and the electromagnetic Hamiltonian $\mathcal{H} = (8\pi)^{-1}(\vec{E}^2 + \vec{B}^2)$. Then, the electric charge is defined through the flux of the electric field as $e = (4\pi)^{-1} \int_{S_R} d^2 S_i E_i$, and analogously for the magnetic charge. We use the convention according to which $e^2 = 4\pi\alpha$, and the electromagnetic Hamiltonian $\mathcal{H} = (2g^2)^{-1}(\vec{E}^2 + \vec{B}^2)$. Then $e = g^{-1} \int_{S_R} d^2 S_i E_i$ while $Q_M = g^{-1} \int_{S_R} d^2 S_i B_i$.

Combining Eqs. (3.3.18), (3.3.17) and (3.3.16) we conclude that

$$E = v \, Q_M + \int d^3x \left\{ \frac{1}{2} \left(\frac{1}{g} B_i^a - D_i \phi^a \right) \left(\frac{1}{g} B_i^a - D_i \phi^a \right) \right\}. \quad (3.3.21)$$

The minimum of the energy functional is attained at

$$\frac{1}{g} B_i^a - D_i \phi^a = 0. \quad (3.3.22)$$

The mass of the field configuration realizing this minimum – the monopole mass –
is obviously equal

$$M_M = \frac{4\pi \, v}{g}. \quad (3.3.23)$$

Thus, the mass of the critical monopole is in one-to-one relation with its magnetic
charge. Equation (3.3.22) is nothing but the Bogomol'nyi equation in the monopole
problem. If it is satisfied, the second-order differential equations of motion are
satisfied too.

3.3.4 Solution of the Bogomol'nyi equation for monopoles

To solve the Bogomol'nyi equations we need to find an appropriate *ansatz* for ϕ^a.
As one sweeps S_R the vector ϕ^a must sweep the group space sphere. The simplest
choice is to identify these two spheres point-by-point,

$$\phi^a = v \frac{x^a}{r} = v n^a, \quad r \to \infty. \quad (3.3.24)$$

where $n^i \equiv x^i / r$. This field configuration obviously belongs to the class with $n = 1$.
The SU(2) group index a got entangled with the coordinate $\vec{x}$. Polyakov proposed
to refer to such fields as "hedgehogs."

Next, observe that finiteness of the monopole energy requires the covariant
derivative $D_i \phi^a$ to fall off faster than $r^{-3/2}$ at large r, cf. Eq. (3.3.13). Since

$$\partial_i \phi^a = v \frac{1}{r} \left\{ \delta^{ai} - n^a n^i \right\} \sim \frac{1}{r} \quad (3.3.25)$$

one must choose A_i^b in such a way as to cancel (3.3.25). It is not difficult to see that

$$A_i^a = \varepsilon^{aij} \frac{1}{r} n^j, \quad r \to \infty. \quad (3.3.26)$$

Then the term $1/r$ is canceled in $D_i \phi^a$.

Equations (3.3.24) and (3.3.26) determine the index structure of the field configuration we are going to deal with. The appropriate *ansatz* is perfectly clear now,

$$\phi^a = v\, n^a\, H(r), \quad A_i^a = \varepsilon^{aij}\, \frac{1}{r}\, n^j\, F(r), \tag{3.3.27}$$

where H and F are functions of r with the boundary conditions

$$H(r) \to 1, \quad F(r) \to 1 \quad \text{at } r \to \infty, \tag{3.3.28}$$

and

$$H(r) \to 0, \quad F(r) \to 0 \quad \text{at } r \to 0. \tag{3.3.29}$$

The boundary condition (3.3.28) is equivalent to Eqs. (3.3.24) and (3.3.26), while the boundary condition (3.3.29) guarantees that our solution is nonsingular at $r \to 0$.

After some straightforward algebra we get

$$B_i^a = \left(\delta^{ai} - n^a n^i\right) \frac{1}{r} F' + n^a n^i \frac{1}{r^2}\left(2F - F^2\right),$$
$$D_i\phi^a = v\left\{\left(\delta^{ai} - n^a n^i\right) \frac{1}{r} H(1 - F) + n^a n^i H'\right\}, \tag{3.3.30}$$

where prime denotes differentiation with respect to r.

Let us return now to the Bogomol'nyi equations (3.3.22). This is a set of nine first-order differential equations. Our *ansatz* has only two unknown functions. The fact that the *ansatz* goes through and we get two scalar equations on two unknown functions from the Bogomol'nyi equations is a highly nontrivial check. Comparing Eqs. (3.3.22) and (3.3.30) we get

$$\frac{1}{g} F' = v\, H(1 - F),$$
$$H' = \frac{1}{g v} \frac{1}{r^2}\left(2F - F^2\right). \tag{3.3.31}$$

The functions H and F are dimensionless. It is convenient to make the radius r dimensionless too. A natural unit of length in the problem at hand is $(gv)^{-1}$. From now on we will measure r in these units,

$$\rho = r\,(gv). \tag{3.3.32}$$

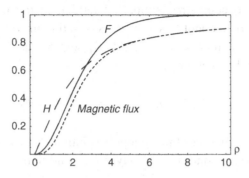

Figure 3.9. The functions F (solid line) and H (long dashes) in the critical monopole solution, vs. ρ. The short-dashed line shows the flux of the magnetic field $\mathcal{B}_i$ (in the units 4π) through the sphere of radius ρ.

The functions H and F are to be considered as functions of ρ, while the prime will denote differentiation over ρ. Then the system (3.3.31) takes the form

$$F' = H(1 - F),$$
$$H' = \frac{1}{\rho^2}\left(2F - F^2\right). \qquad (3.3.33)$$

These equations have known analytical solutions,

$$F = 1 - \frac{\rho}{\sinh\rho},$$
$$H = \frac{\cosh\rho}{\sinh\rho} - \frac{1}{\rho}. \qquad (3.3.34)$$

At large ρ the functions H and F tend to unity (cf. Eq. (3.3.28)) while at $\rho \to 0$

$$F = O(\rho^2), \quad H = O(\rho).$$

They are plotted in Fig. 3.9. Calculating the flux of the magnetic field through the large sphere we verify that for the solution at hand $Q_M = 4\pi/g$.

3.3.5 Collective coordinates (moduli)

The monopole solution presented in the previous section breaks a number of valid symmetries of the theory, for instance, translational invariance. As usual, the symmetries are restored after the introduction of the collective coordinates (moduli), which convert a given solution into a family of solutions.

Our first task is to count the number of moduli in the monopole problem. A straightforward way to count this number is counting linearly independent zero modes. To this end, one represents the fields A_μ and ϕ as a sum of the monopole background plus small deviations,

$$A_\mu^a = A_\mu^{a(0)} + a_\mu^a, \quad \phi^a = \phi^{a(0)} + (\delta\phi)^a, \tag{3.3.35}$$

where the superscript (0) marks the monopole solution. At this point it is necessary to impose a gauge-fixing condition. A convenient condition is

$$\frac{1}{g} D_i a_i^a - \varepsilon^{abc} \phi^b (\delta\phi)^c = 0, \tag{3.3.36}$$

where the covariant derivative in the first term contains only the background field.

Substituting the decomposition (3.3.35) in the Lagrangian one finds the quadratic form for $\{a, (\delta\phi)\}$, and determines the zero modes of this form (subject to the condition (3.3.36)).

We will not trace this procedure in detail, referring the reader to the original literature [109]. Instead, we suggest a simple heuristic consideration.

Let us ask ourselves what are the valid symmetries of the model at hand? They are: (i) three translations; (ii) three spatial rotations; (iii) three rotations in the SU(2) group. Not all these symmetries are independent. It is not difficult to check that the spatial rotations are equivalent to the SU(2) group rotations for the monopole solution. Thus, we should not count them independently. This leaves us with six symmetry transformations.

One should not forget, however, that two of those six act non-trivially in the "trivial vacuum." Indeed, the latter is characterized by the condensate (3.3.4). While rotations around the third axis in the isospace leave the condensate intact (see Eq. (3.3.5)), the rotations around the first and second axes do not. Thus, the number of moduli in the monopole problem is $6 - 2 = 4$. These four collective coordinates have a very transparent physical interpretation. Three of them correspond to translations. They are introduced in the solution through the substitution

$$\vec{x} \to \vec{x} - \vec{x}_0. \tag{3.3.37}$$

The vector $\vec{x}_0$ now plays the role of the monopole center. The unit vector $\vec{n}$ is now defined as $\vec{n} = (\vec{x} - \vec{x}_0)/|\vec{x} - \vec{x}_0|$.

The fourth collective coordinate is related to the unbroken U(1) symmetry of the model. This is the rotation around the direction of alignment of the field ϕ. In the "trivial vacuum" ϕ^a is aligned along the third axis. The monopole generalization of Eq. (3.3.5) is

$$
\begin{aligned}
A^{(0)} &\to U^{-1} A^{(0)} U - i U^{-1} \partial U, \\
\phi^{(0)} &\to U^{-1} \phi^{(0)} U = \phi^{(0)}, \\
U &= \exp \left\{ i \alpha \phi^{(0)} / v \right\},
\end{aligned}
\tag{3.3.38}
$$

where the fields $A^{(0)}$ and $\phi^{(0)}$ are understood here in the matrix form,

$$
A^{(0)} = A^{a(0)} (\tau^a / 2), \quad \phi^{(0)} = \phi^{a(0)} (\tau^a / 2).
$$

Unlike the vacuum field, which is not changed under (3.3.5), the monopole solution for the vector field changes its form. The change looks as a gauge transformation. Note, however, that the gauge matrix U does not tend to unity at $r \to \infty$. Thus, this transformation is in fact a global U(1) rotation. The physical meaning of the collective coordinate α will become clear shortly. Now let us note that (i) for small α Eq. (3.3.38) reduces to

$$
\delta A_i^a = \alpha \frac{1}{v} (D_i \phi^{(0)})^a, \quad \delta \phi = 0,
\tag{3.3.39}
$$

and this is compatible with the gauge condition (3.3.36); (ii) the variable α is compact, since the points α and $\alpha + 2\pi$ can be identified (the transformation of $A^{(0)}$ is identically the same for α and $\alpha + 2\pi$). In other words, α is an angle variable.

Having identified all four moduli relevant to the problem we can proceed to the quasiclassical quantization. The task is to obtain quantum mechanics of the moduli. Let us start from the monopole center coordinate $\vec{x}_0$. To this end, as usual, we assume that $\vec{x}_0$ weakly depends on time t, so that the only time dependence of the solution enters through $\vec{x}_0(t)$. The time dependence is important only in time derivatives, so that the quantum-mechanical Lagrangian of the moduli can be obtained from the following expression:

$$
\begin{aligned}
\mathcal{L}_{\text{QM}} = - M_M + \frac{1}{2} (\dot{x}_0)_k (\dot{x}_0)_j \int d^3x \Bigg\{ & \left[\frac{1}{g} F_{ik}^{a(0)} \right] \left[\frac{1}{g} F_{ij}^{a(0)} \right] \\
& + \left[D_k \phi^{a(0)} \right] \left[D_j \phi^{a(0)} \right] \Bigg\},
\end{aligned}
\tag{3.3.40}
$$

where $\partial_k A$ and $\partial_k \phi$ where supplemented by appropriate gauge transformations to satisfy the gauge condition (3.3.36).

Averaging over the angular orientations of $\vec{x}$ yields

$$\mathcal{L}_{\mathrm{QM}} = -M_M + \frac{1}{2}(\dot{\vec{x}}_0)^2 \int d^3x \left\{ \frac{2}{3}\frac{1}{g^2} B_i^{a(0)} B_i^{a(0)} + \frac{1}{3} D_i\phi^{a(0)} D_i\phi^{a(0)} \right\}$$

$$= -M_M + \frac{M_M}{2}(\dot{\vec{x}}_0)^2. \tag{3.3.41}$$

This last result readily follows if one combines Eqs. (3.3.13) and (3.3.22). Of course, this final answer could have been guessed from the very beginning since this is nothing but the Lagrangian describing free non-relativistic motion of a particle of mass M_M endowed with the coordinate $\vec{x}_0$.

Now, having tested the method in the case where the answer was obvious, let us apply it to the fourth collective coordinate α. Using Eq. (3.3.39) we get

$$\mathcal{L}_{\alpha\mathrm{QM}} = \frac{1}{2}\frac{M_M}{M_W^2}\dot{\alpha}^2, \tag{3.3.42}$$

or, equivalently,

$$\mathcal{H}_\alpha = \frac{1}{2}\frac{M_W^2}{M_M} p_\alpha^2, \quad p_\alpha \equiv -i\frac{d}{d\alpha}, \tag{3.3.43}$$

where $\mathcal{H}_\alpha$ is the part of the Hamiltonian relevant to α. The full quantum-mechanical Hamiltonian describing the moduli dynamics is, thus,

$$\mathcal{H} = M_M + \frac{p^2}{2M_M} + \frac{1}{2}\frac{M_W^2}{M_M} p_\alpha^2, \quad p \equiv -i\frac{d}{dx_0}. \tag{3.3.44}$$

It describes free motion of a spinless particle endowed with an internal (compact) variable α. While the spatial part of $\mathcal{H}$ does not raise any questions, the α dynamics deserves additional discussion.

The α motion is free, but one should not forget that α is an angle. Because of the 2π periodicity, the corresponding wave functions must have the form

$$\Psi(\alpha) = e^{ik\alpha}, \tag{3.3.45}$$

where k is an integer, $k = 0, \pm 1, \pm 2, \ldots$. Strictly speaking, only the ground state, $k = 0$, describes the monopole – a particle with the magnetic charge $4\pi/g$ and vanishing electric charge. Excitations with $k \neq 0$ correspond to a particle with the magnetic charge $4\pi/g$ *and* the electric charge kg, the dyon.

To see that this is indeed the case, let us note that for $k \neq 0$ the expectation value of p_α is k and, hence, the expectation value of $\dot{\alpha} = (M_W^2/M_M)\, p_\alpha$ is $M_W^2 k/M_M$.

Moreover, let us define a gauge-invariant electric field $\mathcal{E}_i$ (analogous to $\mathcal{B}_i$ of Eq. (3.3.17)) as

$$\mathcal{E}_i \equiv \frac{1}{v} E_i^a \phi^a = \frac{1}{v} \phi^{a(0)} \dot{A}_i^{a(0)} = \frac{1}{v^2} \dot{\alpha} \, \phi^{a(0)} \, (D_i \phi^{a(0)}). \tag{3.3.46}$$

Since for the critical monopole $D_i \phi^{a(0)} = (1/g) B_i^{a(0)}$ we see that

$$\mathcal{E}_i = \dot{\alpha} \, \frac{1}{M_W} \mathcal{B}_i, \tag{3.3.47}$$

and the flux of the gauge-invariant electric field over the large sphere is

$$\frac{1}{g} \int_{S_R} d^2 S_i \, \mathcal{E}_i = \frac{M_W^2 k}{M_M} \frac{1}{M_W} \frac{1}{g} \int_{S_R} d^2 S_i \, \mathcal{B}_i \tag{3.3.48}$$

where we replaced $\dot{\alpha}$ by its expectation value. Thus, the flux of the electric field reduces to

$$\frac{1}{g} \int_{S_R} d^2 S_i \, \mathcal{E}_i = kg, \tag{3.3.49}$$

which proves the above assertion of the electric charge kg.

It is interesting to note that the mass of the dyon can be written as

$$M_D = M_M + \frac{1}{2} \frac{M_W^2}{M_M} k^2 \approx \sqrt{M_M^2 + M_W^2 \, k^2} = v \sqrt{Q_M^2 + Q_E^2}. \tag{3.3.50}$$

Although from our derivation it might seem that the square root formula is approximate, in fact, the prediction for the dyon mass $M_D = v(Q_M^2 + Q_E^2)^{1/2}$ is exact; it follows from the BPS saturation and the central charges in $\mathcal{N} = 2$ model (see Chapter 2).

Magnetic monopoles were introduced in theory by Dirac in 1931 [108]. He considered macroscopic electrodynamics and derived a self-consistency condition for the product of the magnetic charge of the monopole Q_M and the elementary electric charge e,[9]

$$Q_M e = 2\pi. \tag{3.3.51}$$

This is known as the Dirac quantization condition. For the 't Hooft–Polyakov monopole we have just derived that $Q_M g = 4\pi$, twice larger than in the Dirac

---

[9] In Dirac's original convention the charge quantization condition is, in fact, $Q_M e = (1/2)$.

quantization condition. Note, however, that g is the electric charge of the W bosons. It is *not* the minimal possible electric charge that can be present in the theory at hand. If quarks in the fundamental (doublet) representation of SU(2) were introduced in the Georgi–Glashow model, their U(1) charge would be $e = g/2$, and the Dirac quantization condition would be satisfied for such elementary charges.

3.3.6 Singular gauge, or how to comb a hedgehog

The *ansatz* (3.3.27) for the monopole solution we used so far is very convenient for revealing a nontrivial topology lying behind this solution, i.e. the fact that $SU(2)/U(1) \sim S_2$ in the group space is mapped onto the spatial S_2. However, it is often useful to gauge-transform it in such a way that the scalar field becomes oriented along the third axis in the color space, $\phi^a \sim \delta^{3a}$, in all space (i.e. at all x), repeating the pattern of the "plane" vacuum (3.3.4). Polyakov suggested to refer to this gauge transformation as "combing the hedgehog." Comparison of Figs. 3.10a and 3.10b shows that this gauge transformation cannot be nonsingular. Indeed, the matrix which combs the hedgehog,

$$U^\dagger \left(n^a \tau^a \right) U = \tau^3, \tag{3.3.52}$$

has the form

$$U = \frac{1}{\sqrt{2}} \left(\sqrt{1 + n^3} + i \, \frac{v^a \tau^a}{\sqrt{1 + n^3}} \right), \tag{3.3.53}$$

where

$$v^a = \varepsilon^{3ab} \, n^b, \quad v^a v^a = 1 - (n^3)^2, \tag{3.3.54}$$

and $\vec{n}$ is the unit vector in the direction of $\vec{x}$. The matrix U is obviously singular at $n^3 = -1$ (see Fig. 3.10). This is a gauge artifact since all physically measurable quantities are nonsingular and well-defined. In the "old" Dirac description of the monopole [110] the singularity of U at $n^3 = -1$ would correspond to the Dirac string.

In the singular gauge the monopole magnetic field at large $|\vec{x}|$ takes the "color-combed" form

$$B_i \to \frac{\tau^3}{2} \frac{n^i}{r^2} = 4\pi \frac{\tau^3}{2} \frac{n^i}{4\pi \, r^2}. \tag{3.3.55}$$

The latter equation naturally implies the same magnetic charge $Q_M = 4\pi/g$, as was derived in Section 3.3.2.

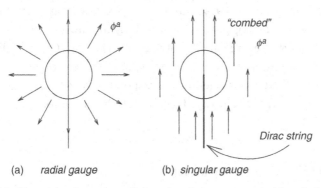

(a) *radial gauge*   (b) *singular gauge*

Figure 3.10. Transition from the radial to singular gauge or combing the hedgehog.

3.3.7 *Monopoles in* SU(N)

Let us now extend the construction presented above from SU(2) to SU(N) [111, 112]. The starting Lagrangian is the same as in Eq. (3.3.1), with the replacement of the structure constants ε^{abc} of SU(2) by the SU(N) structure constants f^{abc}. The potential of the scalar-field self-interaction can be of a more general form than in Eq. (3.3.1). Details of this potential are unimportant for our purposes since in the critical limit the potential tends to zero; its only role is to fix the vacuum value of the field ϕ at infinity.

Recall that all generators of the Lie algebra can be always divided into two groups – the Cartan generators H_i, which all commute with each other, and a set of raising (lowering) operators E_α,

$$E_\alpha^\dagger = E_{-\alpha}. \tag{3.3.56}$$

For SU(N) – and we will not discuss other groups – there are $N-1$ Cartan generators which can be chosen as

$$H^1 = \frac{1}{2}\,\text{diag}\,\{1, -1, 0, \ldots, 0\},$$

$$H^2 = \frac{1}{2\sqrt{3}}\,\text{diag}\,\{1, 1, -2, 0, \ldots, 0\},$$

$$\cdots$$

$$H^m = \frac{1}{\sqrt{2m(m+1)}}\,\text{diag}\,\{1, 1, 1, \ldots, -m, \ldots, 0\},$$

$$\cdots \tag{3.3.57}$$

$$H^{N-1} = \frac{1}{\sqrt{2N(N-1)}}\,\text{diag}\,\{1, 1, 1, \ldots, 1, -(N-1)\},$$

$N(N-1)/2$ raising generators E_α, and $N(N-1)/2$ lowering generators $E_{-\alpha}$. The Cartan generators are analogs of $\tau_3/2$ while $E_{\pm\alpha}$ are analogs of $\tau_\pm/2$. Moreover, $N(N-1)$ vectors α, $-\alpha$ are called root vectors. They are $(N-1)$-dimensional.

By making an appropriate choice of basis, any element of SU(N) algebra can be brought to the Cartan subalgebra. Correspondingly, the vacuum value of the (matrix) field $\phi \equiv \phi^a T^a$ can always be chosen to be of the form

$$\phi_{\mathrm{vac}} = h\,H, \tag{3.3.58}$$

where h is an $(N-1)$-component vector,

$$h = \{h_1, h_2, \ldots, h_{N-1}\}. \tag{3.3.59}$$

For simplicity we will assume that for all simple roots $h\,\gamma > 0$ (otherwise, we will just change the condition defining positive roots to meet this constraint).

Depending on the form of the self-interaction potential distinct patterns of gauge symmetry breaking can take place. We will discuss here the case when the gauge symmetry is maximally broken,

$$\mathrm{SU}(N) \to \mathrm{U}(1)^{N-1}. \tag{3.3.60}$$

The unbroken subgroup is Abelian. This situation is general. In special cases, when h is orthogonal to α^m for some m (or a set of m's) the unbroken subgroup will contain non-Abelian factors, as will be explained momentarily. These cases will not be considered here.

The topological argument proving the existence of a variety of topologically stable monopoles in the above set-up parallels that of Section 3.3.2, except that Eq. (3.3.12) is replaced by

$$\pi_2\left(\mathrm{SU}(N)/\mathrm{U}(1)^{N-1}\right) = \pi_1\left(\mathrm{U}(1)^{N-1}\right) = Z^{N-1}. \tag{3.3.61}$$

There are $N-1$ independent windings in the SU(N) case.

The gauge field A_μ (in the matrix form, $A_\mu \equiv A_\mu^a T^a$) can be represented as

$$A_\mu^a T^a = \sum_{m=1}^{N-1} A_\mu^m H^m + \sum_\alpha A_\mu^\alpha E_\alpha, \tag{3.3.62}$$

where A_μ^m's ($m = 1, \ldots, N-1$) can be viewed as "photons," while A_μ^α's as "W bosons." The mass terms are obtained from the term

$$\mathrm{Tr}\left([A_\mu, \phi]\right)^2$$

in the Lagrangian. Substituting here Eqs. (3.3.58) and (3.3.62) it is easy to see that the W-boson masses are

$$M\alpha = g\,h\alpha. \tag{3.3.63}$$

$N - 1$ massive bosons corresponding to simple roots γ play a special role: they can be thought of as fundamental, in the sense that the quantum numbers and masses of all other W bosons can be obtained as linear combinations (with non-negative integer coefficients) of those of the fundamental W bosons. With regards to the masses this is immediately seen from Eq. (3.3.63) in conjunction with

$$\alpha = \sum_{\gamma} k_{\gamma}\,\gamma. \tag{3.3.64}$$

Construction of SU(N) monopoles reduces, in essence, to that of a SU(2) monopole through various embeddings of SU(2) in SU(N). Note that each simple root γ defines an SU(2) subgroup[10] of SU(N) with the following three generators:

$$t^1 = \frac{1}{\sqrt{2}}\left(E_{\gamma} + E_{-\gamma}\right),$$
$$t^2 = \frac{1}{\sqrt{2}\,i}\left(E_{\gamma} - E_{-\gamma}\right),$$
$$t^3 = \gamma\,H, \tag{3.3.65}$$

with the standard algebra $[t^i, t^j] = i\varepsilon^{ijk}\,t^k$. If the basic SU(2) monopole solution corresponding to the Higgs vacuum expectation value v is denoted as $\{\phi^a(r;\,v), A_i^a(r;\,v)\}$, see Eq. (3.3.27), the construction of a specific SU(N) monopole proceeds in three steps: (i) choose a simple root γ; (ii) decompose the vector h in two components, parallel and perpendicular with respect to γ,

$$h = h_{\parallel} + h_{\perp},$$
$$h_{\parallel} = \tilde{v}\gamma, \quad h_{\perp}\gamma = 0,$$
$$\tilde{v} \equiv \gamma\,h > 0; \tag{3.3.66}$$

(iii) replace $A_i^a(r;\,v)$ by $A_i^a(r;\,\tilde{v})$ and add a covariantly constant term to the field $\phi^a(r;\,\tilde{v})$ to ensure that at $r \to \infty$ it has the correct asymptotic behavior, namely, $2\,\mathrm{Tr}\,\phi^2 = h^2$. Algebraically the SU($N$) monopole solution takes the form

$$\phi = \phi^a(r;\,\tilde{v})\,t^a + h_{\perp}H, \quad A_i = A_i^a(r;\,\tilde{v})\,t^a. \tag{3.3.67}$$

---

[10] Generally speaking, each root α defines an SU(2) subalgebra according to Eq. (3.3.65), but we will deal only with the simple roots for reasons which will become clear momentarily.

Note that the mass of the corresponding W boson $M_\gamma = g\tilde{v}$, in full parallel with the SU(2) monopole.

It is instructive to verify that (3.3.67) satisfies the BPS equation (3.3.22). To this end it is sufficient to note that $[h_\perp H, A_i] = 0$, which in turn implies

$$D_i (h_\perp H) = 0.$$

What remains to be done? We must analyze the magnetic charges of the SU(N) monopoles and their masses. In the singular gauge (Section 3.3.6) the Higgs field is aligned in the Cartan subalgebra, $\phi \sim h H$. The magnetic field at large distances from the monopole core, being commutative with ϕ, also lies in the Cartan subalgebra. In fact, from Eq. (3.3.65) we infer that combing of the SU(N) monopole leads to

$$B_i \to 4\pi \, \gamma H \, \frac{n^i}{4\pi \, r^2}, \tag{3.3.68}$$

which implies, in turn, that the set of $N-1$ magnetic charges of the SU(N) monopole is given by the components of the $(N-1)$-vector

$$Q_M = \frac{4\pi}{g} \gamma. \tag{3.3.69}$$

Of course, the very same result is obtained in a gauge invariant manner from a defining formula

$$2\text{Tr} (B_i \phi) \xrightarrow{r \to \infty} (Q_M h) \frac{g}{4\pi} \frac{n_i}{r^2}. \tag{3.3.70}$$

Equation (3.3.15) implies that the mass of this monopole is

$$M_{M\gamma} = Q_M h = \frac{4\pi \, \tilde{v}}{g}, \tag{3.3.71}$$

to be compared with the mass of the corresponding W bosons,

$$M_\gamma = g\gamma h = g\tilde{v}, \tag{3.3.72}$$

in perfect parallel with the SU(2) monopole results of Section 3.3.3. The general magnetic charge quantization condition takes the form

$$\exp\{ig \, Q_M H\} = 1. \tag{3.3.73}$$

Let us ask ourselves what happens if one builds monopoles on non-simple roots. Such solutions are in fact composite: they consist of the basic "simple-root" monopoles – the masses and quantum numbers (magnetic charges) of the composite monopoles can be obtained by summing up the masses and quantum numbers of the basic monopoles, according to Eq. (3.3.64).

3.3.8 The θ term induces a fractional electric charge for the monopole (the Witten effect)

There is a P- and T-odd term, the θ term, which can be added to the Lagrangian for the Yang–Mills theory without spoiling renormalizability. It is given by

$$\mathcal{L}_\theta = \frac{\theta}{32\pi^2} F^a_{\mu\nu} \tilde{F}^{a\mu\nu} = -\frac{\theta}{8\pi^2} \vec{E}^a \cdot \vec{B}^a. \tag{3.3.74}$$

This interaction violates P and CP but not C. As is well known, this term is a surface term and does not affect the classical equations of motion. There is, however, a θ dependence in instanton effects which involve nontrivial long-range behavior of the gauge fields. As was realized by Witten [113], in the presence of magnetic monopoles θ also has a nontrivial effect, it shifts the allowed values of electric charge in the monopole sector of the theory.

Since the equations of motions do not change, the monopole solution obtained above stays intact. What changes is the effective quantum-mechanical Lagrangian. As usual, we assume an adiabatic time dependence of moduli. In the case at hand we must replace the constant phase modulus α by $\alpha(t)$. This generates the electric field

$$E^a_i = \dot{\alpha} \left(\delta A^a_i / \delta \alpha \right) = \frac{\dot{\alpha}}{v} \left(D_i \phi^{(0)} \right)^a,$$

where Eq. (3.3.39) is used. The magnetic field does not change, and can be expressed through $\left(D_i \phi^{(0)} \right)^a$ using Eq. (3.3.22). As a result, the quantum-mechanical Lagrangian for α acquires a full derivative term,

$$\mathcal{L}_{\alpha\mathrm{QM}} = \frac{1}{2\mu} \dot{\alpha}^2 - \frac{\theta}{2\pi} \dot{\alpha}, \quad \mu = \frac{M_W^2}{M_M}. \tag{3.3.75}$$

This changes the expression for the canonic momentum conjugated to α. If previously p_α was $\dot{\alpha}/\mu$, now

$$p_\alpha = \frac{\dot{\alpha}}{\mu} - \frac{\theta}{2\pi}. \tag{3.3.76}$$

Correspondingly,

$$\dot{\alpha} = \mu \left(p_\alpha + \frac{\theta}{2\pi} \right). \tag{3.3.77}$$

From Sect. 3.3.5 we know that the electric charge of the field configuration at hand is (see Eq. (3.3.49))

$$Q_E = \frac{1}{M_W g} \langle \dot{\alpha} \rangle \int_{S_R} d^2 S_i \, \mathcal{B}_i.$$ (3.3.78)

Substituting Eq. (3.3.77) and $\langle p_\alpha \rangle = k$ we arrive at

$$Q_E = \left(k + \frac{\theta}{2\pi} \right) g.$$ (3.3.79)

We see that at $\theta \neq 0$ the electric charge of the dyon is non-integer. As θ changes from zero to the physically equivalent point $\theta = 2\pi$ the dyon charges shift by one unit. The dyon spectrum as a whole remains intact.

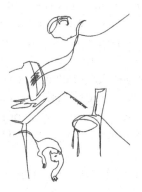

3.4 Monopoles and fermions

The critical 't Hooft–Polyakov monopoles we have just discussed can be embedded in $\mathcal{N} = 2$ super-Yang–Mills. There are no $\mathcal{N} = 1$ models with the 't Hooft–Polyakov monopoles (albeit $\mathcal{N} = 1$ theories supporting confined monopoles are found [104]). The minimal model with the BPS-saturated 't Hooft–Polyakov monopole is the $\mathcal{N} = 2$ generalization of supersymmetric gluodynamics, with the gauge group SU(2). In terms of $\mathcal{N} = 1$ superfields it contains one vector superfield in the adjoint describing gluon and gluino, plus one chiral superfield in the adjoint describing a scalar $\mathcal{N} = 2$ superpartner for gluon and a Weyl spinor, an $\mathcal{N} = 2$ superpartner for gluino.

The couplings of the fermion fields to the boson fields are of a special form, they are fixed by $\mathcal{N} = 2$ supersymmetry. In this section we will first present the Lagrangian of $\mathcal{N} = 2$ supersymmetric gluodynamics, including the part with the adjoint fermions, and then consider effects due to the adjoint fermions. We conclude

Section 3.4 with a comment on fermions in the fundamental representation in the monopole background.

3.4.1 $\mathcal{N} = 2$ super-Yang–Mills (without matter)

Two $\mathcal{N} = 1$ superfields are used to build the model,

$$W_\alpha = i \left(\lambda_\alpha + i\theta_\alpha D - \theta^\beta F_{\alpha\beta} - i\theta^2 D_{\alpha\dot\alpha} \bar\lambda^{\dot\alpha} \right), \qquad (3.4.1)$$

and

$$\mathcal{A} = a + \sqrt{2}\,\psi\,\theta + \theta^2 F. \qquad (3.4.2)$$

Here the notation is spinorial, and all fields are in the adjoint representation of SU(2). The corresponding generators are

$$\left(T^a \right)_{bd} = i\,\varepsilon_{bad}. \qquad (3.4.3)$$

The Lagrangian contains kinetic terms and their supergeneralizations. In components

$$
\begin{aligned}
\mathcal{L} = \frac{1}{g^2} \Big\{ & -\frac{1}{4} F^{a\,\mu\nu} F^a_{\mu\nu} + \lambda^{\alpha,a}\, i\, D_{\alpha\dot\alpha}\, \bar\lambda^{\dot\alpha,a} + \frac{1}{2}\, D^a D^a \\
& + \psi^{\alpha,a}\, i\, D_{\alpha\dot\alpha}\, \bar\psi^{\dot\alpha,a} + (D^\mu\,\bar a)(D_\mu\,a) \\
& - \sqrt{2}\,\varepsilon_{abc} \left(\bar a^a\, \lambda^{\alpha,b}\, \psi^c_\alpha + a^a\, \bar\lambda^b_{\dot\alpha}\, \bar\psi^{\dot\alpha,c} \right) - i\,\varepsilon_{abc}\, D^a\, \bar a^b\, a^c \Big\}.
\end{aligned} \qquad (3.4.4)
$$

As usual, the D field is auxiliary and can be eliminated via the equation of motion,

$$D^a = i\,\varepsilon_{abc}\,\bar a^b a^c. \qquad (3.4.5)$$

There is a flat direction: if the field a is real all D terms vanish. If a is chosen to be purely real or purely imaginary and the fermion fields ignored we obviously return to the Georgi–Glashow model discussed above.

Let us perform the Bogomol'nyi completion of the bosonic part of the Lagrangian (3.4.4) for static field configurations. Neglecting all time derivatives and, as usual, setting $A_0 = 0$, one can write the energy functional as follows:

$$
\begin{aligned}
\mathcal{E} = \sum_{i=1,2,3;\, a=1,2,3} \int d^3x & \left[\frac{1}{\sqrt{2}g} F_i^{*a} \pm \frac{1}{g} D_i a^a \right]^2 \\
& \mp \frac{\sqrt{2}}{g^2} \int d^3x\, \partial_i \left(F_i^{*a}\, a^a \right),
\end{aligned} \qquad (3.4.6)
$$

where

$$F_m^* = \frac{1}{2} \varepsilon_{mnk} F_{nk},$$

and the square of the D term (3.4.5) is omitted – the D term vanishes provided a is real, which we will assume. This assumption also allows us to replace the absolute value in the first line by the square brackets. The term in the second line can be written as an integral over a large sphere,

$$\frac{\sqrt{2}}{g^2} \int d^3x \, \partial_i \left(F_i^{*a} \, a^a \right) = \frac{\sqrt{2}}{g^2} \int dS_i \left(a^a \, F_i^{*a} \right). \tag{3.4.7}$$

The Bogomol'nyi equations for the monopole are

$$F_i^{*a} \pm \sqrt{2} D_i \, a^a = 0. \tag{3.4.8}$$

This coincides with Eq. (3.3.22) in the Georgi–Glashow model, up to a normalization. (The field a is complex, generally speaking, and its kinetic term is normalized differently.) If the Bogomol'nyi equations are satisfied, the monopole mass is determined by the surface term (classically). Assuming that in the "flat" vacuum a^a is aligned along the third direction and taking into account that in our normalization the magnetic flux is 4π we get

$$M_M = \frac{\sqrt{2} \, a_{\text{vac}}^3}{g^2} \, 4\pi, \tag{3.4.9}$$

where – we recall – a_{vac}^3 is assumed to be positive. This is in full agreement with Eq. (3.3.23).

3.4.2 Supercurrents and the monopole central charge

The general classification of central charges in $\mathcal{N} = 2$ theories in four dimensions is presented in Section 2.3.3. Here we will briefly discuss the Lorentz-scalar central charge Z in the theory (3.4.4). It is this central charge that is saturated by critical monopoles.

The model, being $\mathcal{N} = 2$, possesses two conserved supercurrents,

$$J_{\alpha\beta\dot\beta}^I = \frac{2}{g^2} \left\{ i F_{\beta\alpha}^a \bar\lambda_{\dot\beta}^a + \varepsilon_{\beta\alpha} D^a \bar\lambda_{\dot\beta}^a + \sqrt{2} \left(D_{\alpha\dot\beta} \bar a^a \right) \psi_\beta^a \right\} + \text{f.d.},$$

$$J_{\alpha\beta\dot\beta}^{II} = \frac{2}{g^2} \left\{ i F_{\beta\alpha}^a \bar\psi_{\dot\beta}^a + \varepsilon_{\beta\alpha} D^a \bar\psi_{\dot\beta}^a - \sqrt{2} \left(D_{\alpha\dot\beta} \bar a^a \right) \lambda_\beta^a \right\} + \text{f.d.}, \tag{3.4.10}$$

where f.d. stands for full derivatives. Both expressions can be combined in one compact formula if we introduce an SU(2) index f ($f = 1, 2$) (to be repeatedly used in Part II) in the following way:

$$\lambda^f = \begin{cases} \lambda, & f = 1 \\ \psi, & f = 2. \end{cases} \tag{3.4.11}$$

Then $\lambda_1 = -\psi$ and $\lambda_2 = \lambda$. The supercurrent takes the form ($f = 1, 2$)

$$J_{\alpha\beta\dot\beta,\,f} = \frac{2}{g^2}\left\{ i F^a_{\beta\alpha}\bar\lambda^a_{\dot\beta,\,f} + \varepsilon_{\beta\alpha}D^a\bar\lambda^a_{\dot\beta,\,f} - \sqrt{2}\left(D_{\alpha\dot\beta}\bar a^a\right)\lambda^a_{\beta,\,f} \right.$$
$$\left. + \frac{\sqrt{2}}{6}\left[\partial_{\alpha\dot\beta}(\lambda_{\beta,\,f}\,\bar a) + \partial_{\beta\dot\beta}(\lambda_{\alpha,\,f}\,\bar a) - 3\varepsilon_{\beta\alpha}\partial^{\gamma}_{\dot\beta}(\lambda_{\gamma,\,f}\,\bar a)\right]\right\}. \tag{3.4.12}$$

Classically the commutator of the corresponding supercharges is

$$\{Q^I_\alpha, Q^{II}_\beta\} = 2\,Z\,\varepsilon_{\alpha\beta} = -\frac{2\sqrt{2}}{g^2}\,\varepsilon_{\alpha\beta}\int d^3x\,\mathrm{div}\left(\bar a^a\left(\vec E^a - i\vec B^a\right)\right)$$
$$= -\frac{2\sqrt{2}}{g^2}\,\varepsilon_{\alpha\beta}\int dS_j\left(\bar a^a\left(E^a_j - i\,B^a_j\right)\right). \tag{3.4.13}$$

Z in Eq. (3.4.13) is sometimes referred to as the monopole central charge. For the BPS-saturated monopoles $M_M = Z$.

Quantum corrections in the monopole central charge and in the mass of the BPS saturated monopoles were first discussed in Refs. [8, 114, 43] two decades ago. The monopole central charge is renormalized at one-loop level. This is obvious due to the fact that the corresponding quantum correction must convert the bare coupling constant in Eq. (3.4.13) into the renormalized one. The fact that the logarithmic renormalizations of the monopole mass and the gauge coupling constant match was established long ago. However, there is a residual non-logarithmic effect which cannot be obtained from Eq. (3.4.13). It was not until 2004 that people realized that the monopole central charge (3.4.13) must be supplemented by an anomalous term [39].

To elucidate the point, let us consider (following [38]) the formula for the monopole/dyon mass obtained in the exact Seiberg–Witten solution [2],

$$M_{n_e, n_m} = \sqrt{2}\left|a\left(n_e - \frac{a_D}{a}n_m\right)\right|, \tag{3.4.14}$$

where $n_{e,m}$ are integer electric and magnetic numbers (we will consider here only a particular case when either $n_e = 0, 1$ or $n_m = 0, 1$) and

$$a_D = i\, a \left(\frac{4\pi}{g_0^2} - \frac{2}{\pi} \ln \frac{M_0}{a} \right). \tag{3.4.15}$$

The subscript 0 is introduced for clarity, it marks the bare charge. The renormalized coupling constant is defined in terms of the ultraviolet parameters as follows:

$$\frac{\partial a_D}{\partial a} \equiv \frac{4\pi i}{g^2}. \tag{3.4.16}$$

Because of the $a \ln a$ dependence, $\partial a_D / \partial a$ differs from a_D / a by a constant (non-logarithmic) term, namely,

$$\frac{a_D}{a} = i \left(\frac{4\pi}{g^2} - \frac{2}{\pi} \right). \tag{3.4.17}$$

Combining Eq. (3.4.14) and (3.4.17) we get

$$M_{n_e, n_m} = \sqrt{2} \left| a \left(n_e - i \left(\frac{4\pi}{g^2} - \frac{2}{\pi} \right) n_m \right) \right|, \tag{3.4.18}$$

This does not match Eq. (3.4.13) in the non-logarithmic part (i.e. the term $2\sqrt{2}\, n_m / \pi$). Since the relative weight of the electric and magnetic parts in Eq. (3.4.13) is fixed to be g^2, the presence of the above non-logarithmic term implies that, in fact, the chiral structure $E_j^a - i\, B_j^a$ obtained at the canonic commutator level cannot be maintained once quantum corrections are switched on. This is a quantum anomaly.

So far no direct calculation of the anomalous contribution in $\{Q_\alpha^I, Q_\beta^{II}\}$ in the operator form has been carried out. However, it is not difficult to reconstruct it indirectly, using Eq. (3.4.18) and a close parallel between $\mathcal{N} = 2$ super-Yang–Mills theory and $\mathcal{N} = 2$ CP($N-1$) model with twisted mass in two dimensions in which a similar problem was solved [34],

$$\left\{ Q_\alpha^I, Q_\beta^{II} \right\}\Big|_{\text{anom}} = 2\, \varepsilon_{\alpha\beta}\, \delta Z_{\text{anom}} = -\, (\varepsilon_{\alpha\beta})\, 2\sqrt{2}\, \frac{1}{4\pi^2} \int dS_j\, \Sigma^j \tag{3.4.19}$$

where

$$\Sigma^j = \frac{i}{2} \frac{\partial}{\partial \bar\theta^{\dot\beta}} \left(\bar{A}^a\, \bar{W}_{\dot\alpha}^a \right) \left(\sigma^j \right)^{\dot\alpha\dot\beta} \Big|_{\bar\theta=0}$$
$$= \bar{a}^a \left(\vec{E}^a + i\, \vec{B}^a \right)^j - \frac{\sqrt{2}}{2}\, \bar\lambda_{\dot\alpha}^a \left(\sigma^j \right)^{\dot\alpha\dot\beta}\, \bar\psi_{\dot\beta}^a, \tag{3.4.20}$$

to be added to Eq. (3.4.13). The (1,0) conversion matrix $(\sigma^j)^{\dot\alpha\dot\beta}$ is defined in Section A.5. Equation (3.4.20) is to be compared with that obtained at the end of Section 4.5.3. We hasten to note that the bifermion term $\bar\lambda\bar\psi$ in δZ_{anom} was calculated in Ref. [39].

In the $SU(N)$ theory we would have $N/8\pi^2$ instead of $1/4\pi^2$ in Eq. (3.4.19).

Adding the canonic and the anomalous terms in $\{Q_\alpha^I, Q_\beta^{II}\}$ together we see that the fluxes generated by color-electric and color-magnetic terms are now shifted, untied from each other, by a non-logarithmic term in the magnetic part. Normalizing to the electric term, $M_W = \sqrt{2}a$, we get for the magnetic term

$$M_M = \sqrt{2}\,a\left(\frac{4\pi}{g^2} - \frac{2}{\pi}\right), \qquad (3.4.21)$$

as it is necessary for the consistency with the exact Seiberg–Witten solution.

3.4.3 Zero modes for adjoint fermions

Equations for the fermion zero modes can be readily derived from the Lagrangian (3.4.4),

$$i D_{\alpha\dot\alpha}\lambda^{\alpha,c} - \sqrt{2}\,\varepsilon_{abc}\,a^a\,\bar\psi_{\dot\alpha}^b = 0$$
$$i D_{\alpha\dot\alpha}\psi^{\alpha,c} + \sqrt{2}\,\varepsilon_{abc}\,a^a\,\bar\lambda_{\dot\alpha}^b = 0, \qquad (3.4.22)$$

plus Hermitean conjugate. After a brief reflection we can get two complex (four real) zero modes.[11] Two of them are obtained if we substitute

$$\lambda^\alpha = F^{\alpha\beta}, \quad \bar\psi_{\dot\alpha} = \sqrt{2}\,D_{\alpha\dot\alpha}\,\bar a. \qquad (3.4.23)$$

The other two solutions correspond to the following substitution:

$$\psi^\alpha = F^{\alpha\beta}, \quad \bar\lambda_{\dot\alpha} = \sqrt{2}\,D_{\alpha\dot\alpha}\,\bar a. \qquad (3.4.24)$$

This result is easy to understand. Our starting theory has eight supercharges. The classical monopole solution is BPS-saturated, implying that four of these eight supercharges annihilate the solution (these are the Bogomol'nyi equations), while the action of the other four supercharges produces the fermion zero modes.

With four real fermion collective coordinates, the monopole supermultiplet is four-dimensional: it includes two bosonic states and two fermionic. (The above counting refers just to monopole, without its antimonopole partner. The antimonopole supermultiplet also includes two bosonic and two fermionic states.) From

---

[11] This means that the monopole is described by two complex fermion collective coordinates, or four real.

the standpoint of $\mathcal{N} = 2$ supersymmetry in four dimensions this is a short multiplet. Hence, the monopole states remain BPS saturated to all orders in perturbation theory (in fact, the criticality of the monopole supermultiplet is valid beyond perturbation theory [2, 3]).

3.4.4 Zero modes for fermions in the fundamental representation

This topic, being related to an interesting phenomenon of charge fractionalization, is marginal for this review. Therefore, we will limit ourselves to a brief comment. The interested reader is referred to [16, 115, 17] for further details. The fermion part of the Lagrangian can be obtained from (3.4.4) with the obvious replacement of the adjoint Dirac fermion by the fundamental one, which we will denote by χ,

$$\mathcal{L} = \frac{1}{g^2} \left\{ -\frac{1}{4} F^{a\,\mu\nu} F^a_{\mu\nu} + \frac{1}{2} (D^\mu \phi)(D_\mu \phi) + \bar{\chi}\, i\, \slashed{D}\chi - \bar{\chi}\, \phi\chi \right\}. \qquad (3.4.25)$$

The Dirac equation then takes the form

$$(i\gamma^\mu D_\mu - \phi)\chi = 0. \qquad (3.4.26)$$

Gamma matrices can be chosen in any representation. The one which is most convenient here is

$$\gamma^0 = \begin{pmatrix} 0 & -i \\ i & 0 \end{pmatrix}, \quad \gamma^i = \begin{pmatrix} -i\sigma^i & 0 \\ 0 & i\sigma^i \end{pmatrix}. \qquad (3.4.27)$$

For the static 't Hooft–Polyakov monopole configuration (with $A_0 = 0$) the zero mode equations reduce to two decoupled equations

$$\slashed{D}\chi^- \equiv (\sigma^i D_i + \phi)\chi^- = 0.$$
$$\slashed{D}^\dagger \chi^+ \equiv (\sigma^i D_i - \phi)\chi^+ = 0, \quad i = 1, 2, 3. \qquad (3.4.28)$$

provided we parametrize $\chi(\vec{x})$ in terms of the following two-component spinors:

$$\chi = \begin{pmatrix} \chi^+ \\ \chi^- \end{pmatrix}. \qquad (3.4.29)$$

Now we can use the Callias theorem [116] which says

$$\dim \ker \slashed{D} - \dim \ker \slashed{D}^\dagger = n_m, \qquad (3.4.30)$$

where n_m is the topological number, $n_m = 1$ for the monopole and $n_m = -1$ for the antimonopole. This implies, in turn, that Eq. (3.4.28) has one complex zero mode, i.e. in the case at hand we characterize the monopole by one complex fermion collective coordinate (and a conjugate, of course). This fact leads to a drastic consequence: the monopole acquires a half-integer electric charge. It becomes a dyon with charge 1/2 even in the absence of the θ term. This phenomenon – the charge fractionalization in the cases with a single complex fermion collective coordinate – is well known in the literature [115, 16, 34, 17] and dates back to Jackiw and Rebbi [117].

3.4.5 The monopole supermultiplet: dimension of the BPS representations

As was first noted by Montonen and Olive [118], all states in $\mathcal{N} = 2$ model – W bosons and monopoles alike – are BPS saturated. This results in the fact that supermultiplets of this model are short. Regular (long) supermultiplet would contain $2^{2\mathcal{N}} = 16$ helicity states, while the short ones contain $2^{\mathcal{N}} = 4$ helicity states – two bosonic and two fermionic. This is in full accord with the fact that the number of the fermion zero modes on the given monopole solution is four, resulting in dim-4 representation of the supersymmetry algebra. If we combine particles and antiparticles together, as is customary in field theory, we will have one Dirac spinor on the fermion side of the supermultiplet. This statement is valid in both cases, the monopole supermultiplet and that of W-bosons.

3.5 More on kinks (in $\mathcal{N} = 2$ CP(1) model)

Kinks in two-dimensional $\mathcal{N} = 2$ CP($N - 1$) models will play a crucial role in our subsequent studies of confined monopoles in Part II of this book (see e.g. Sections 4.4.1, 4.4.3, 4.4.4, and 4.5). Here we will review basic features of such kinks using CP(1) as the simplest example.

The Lagrangian of the CP(1) model with the twisted mass has the form [32]

$$\mathcal{L}_{CP(1)} = G \left\{ \partial_\mu \phi^\dagger \, \partial^\mu \phi - |m|^2 \phi^\dagger \phi + \frac{i}{2} (\psi_L^\dagger \overset{\leftrightarrow}{\partial_R} \psi_L + \psi_R^\dagger \overset{\leftrightarrow}{\partial_L} \psi_R) \right.$$

$$- i \frac{1 - \phi^\dagger \phi}{\chi} (m \, \psi_L^\dagger \psi_R + \bar{m} \psi_R^\dagger \psi_L)$$

$$- \frac{i}{\chi} [\psi_L^\dagger \psi_L (\phi^\dagger \overset{\leftrightarrow}{\partial_R} \phi) + \psi_R^\dagger \psi_R (\phi^\dagger \overset{\leftrightarrow}{\partial_L} \phi)]$$

$$\left. - \frac{2}{\chi^2} \psi_L^\dagger \psi_L \psi_R^\dagger \psi_R \right\} + \frac{i g^2 \theta}{4\pi} G \, \varepsilon^{\mu\nu} \partial_\mu \phi^\dagger \, \partial_\nu \phi, \tag{3.5.1}$$

where

$$\partial_L = \frac{\partial}{\partial t} + \frac{\partial}{\partial z}, \quad \partial_R = \frac{\partial}{\partial t} - \frac{\partial}{\partial z}, \tag{3.5.2}$$

and

$$G = \frac{2}{g^2 \chi^2}, \quad \chi = 1 + \phi \phi^\dagger. \tag{3.5.3}$$

Moreover, m is a complex mass parameter and θ is the vacuum angle. The above Lagrangian has an obvious U(1) symmetry. At $m = 0$ it describes the $\mathcal{N} = 2$ supergeneralization of the σ model on the sphere S_2 (see Appendix B). The metric of the sphere G is chosen in the Fubini–Study form.

It is not difficult to derive the supercurrent,

$$J_\alpha^\mu = \sqrt{2} \, G [\partial_\nu \phi^\dagger \gamma^\nu \gamma^\mu \psi + i \phi^\dagger \gamma^\mu \mu \, \psi]_\alpha, \tag{3.5.4}$$

where

$$\mu = m \frac{1 + \gamma_5}{2} + \bar{m} \frac{1 - \gamma_5}{2}. \tag{3.5.5}$$

The superalgebra is centrally extended, as in Eq. (2.3.4) with $Z' = 0$ and

$$Z = m q_{U(1)} - i \int dz \, \partial_z \left\{ m \, D - \frac{1}{2\pi} (m g_0^2 D - i \, R \, \psi_R^\dagger \psi_L) \right\}, \tag{3.5.6}$$

where

$$R = \frac{2}{\chi^2}, \quad D = \frac{2}{g^2} \frac{\phi^\dagger \phi}{\chi}, \tag{3.5.7}$$

and $q_{U(1)}$ is the Noether charge corresponding to the U(1) current

$$\mathcal{J}_\mu = G \left[\phi^\dagger i \overset{\leftrightarrow}{\partial_\mu} \phi + \bar{\psi} \gamma_\mu (\psi + \Gamma \phi \, \psi) \right], \quad \Gamma = -2 \frac{\phi^\dagger}{\chi}. \tag{3.5.8}$$

Thus, the term $m\,q_{U(1)}$ in Z represents the Noether charge while the integral represents the topological charge.

The first two terms are classical, while the term in parentheses is the quantum anomaly derived in [33, 34].[12] At large $|m|$ the theory is at weak coupling.

3.5.1 BPS solitons at the classical level

The U(1) invariant scalar potential term

$$V = |m|^2 \, G \, \bar{\phi}\phi \qquad (3.5.9)$$

lifts the vacuum degeneracy leaving us with two discrete vacua: at the south and north poles of the sphere (Fig. 3.11) i.e. $\phi = 0$ and $\phi = \infty$.

The kink solutions interpolate between these two vacua. Let us focus, for definiteness, on the kink with the boundary conditions

$$\phi \to 0 \quad \text{at} \quad z \to -\infty, \quad \phi \to \infty \quad \text{at} \quad z \to \infty. \qquad (3.5.10)$$

Consider the following linear combinations of supercharges

$$\mathcal{Q} = Q_R - i\,e^{-i\beta}\,Q_L, \quad \bar{\mathcal{Q}} = \bar{Q}_R + i\,e^{i\beta}\,\bar{Q}_L, \qquad (3.5.11)$$

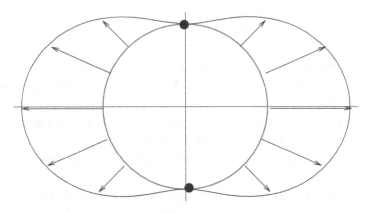

Figure 3.11. Meridian slice of the target space sphere (thick solid line). Arrows present the scalar potential in (3.5.1), their length being the strength of the potential. Two vacua of the model are denoted by closed circles.

[12] In the first of these papers only the bifermion part of the anomaly was obtained. The full anomalous term in the central charge (3.5.6) was found in [34]; later it was confirmed in [119].

where β is the argument of the mass parameter,

$$m = |m|\, e^{i\beta}. \tag{3.5.12}$$

Then

$$\{Q\bar{Q}\} = 2H - 2Z, \quad \{QQ\} = \{\bar{Q}\bar{Q}\} = 0. \tag{3.5.13}$$

Now, let us require Q and $\bar{Q}$ to vanish on the classical solution. Since for static field configurations

$$Q = -\left(\partial_z\bar{\phi} - |m|\bar{\phi}\right)\left(\Psi_R + i e^{-i\beta}\Psi_L\right),$$

the vanishing of these two supercharges implies

$$\partial_z\bar{\phi} = |m|\bar{\phi} \quad \text{or} \quad \partial_z\phi = |m|\phi. \tag{3.5.14}$$

This is the BPS equation in the CP(1) model with the twisted mass.

The BPS equation (3.5.14) has a number of peculiarities compared to those in more familiar Wess–Zumino models. The most important feature is its complexification, i.e. the fact that Eq. (3.5.14) is holomorphic in ϕ.

The solution of this equation is, of course, trivial, and can be written as

$$\phi(z) = e^{|m|(z-z_0)-i\alpha}. \tag{3.5.15}$$

Here z_0 is the kink center while α is an arbitrary phase. In fact, these two parameters enter only in the combination $|m|z_0 + i\alpha$. We see that the notion of the kink center also gets complexified.

The physical meaning of the modulus α is obvious: there is a continuous family of solitons interpolating between the north and south poles of the target space sphere. This is due to U(1) symmetry. The soliton trajectory can follow any meridian (Fig. 3.12).

Equation (3.5.6) for the central charge implies that classically the kink mass is

$$M_0 = |m|\,(D(\infty) - D(0)) = \frac{2|m|}{g^2}. \tag{3.5.16}$$

(The subscript 0 emphasizes that this result is obtained at the classical level.) Quantum corrections will be considered shortly.

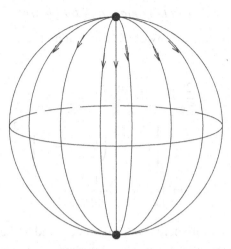

Figure 3.12. The soliton solution family. The collective coordinate α in Eq. (3.5.15) spans the interval $0 \leq \alpha \leq 2\pi$. For given α the soliton trajectory on the target space sphere follows a meridian, so that when α varies from 0 to 2π all meridians are covered.

3.5.2 Quantization of the bosonic moduli

To carry out conventional quasiclassical quantization we, as usual, assume the moduli z_0 and α in Eq. (3.5.15) to be (weakly) time-dependent, substitute (3.5.15) in the bosonic part of the Lagrangian (3.5.1), integrate over z and arrive at

$$\mathcal{L}_{QM} = -M_0 + \frac{M_0}{2}\dot{z}_0^2 + \left\{ \frac{1}{g^2|m|}\dot{\alpha}^2 - \frac{\theta}{2\pi}\dot{\alpha} \right\}. \qquad (3.5.17)$$

The first term is the classical kink mass, the second describes free motion of the kink along the z axis. The term in the braces is most interesting. The variable α is compact. Its very existence is related to the exact U(1) symmetry of the model. The energy spectrum corresponding to α dynamics is quantized. It is not difficult to see that

$$E_{[\alpha]} = \frac{g^2|m|}{4}\,q_{U(1)}^2, \qquad (3.5.18)$$

where $q_{U(1)}$ is the U(1) charge of the soliton,

$$q_{U(1)} = k + \frac{\theta}{2\pi}, \quad k = \text{ an integer.} \qquad (3.5.19)$$

Here we see the Witten phenomenon at work, analogously to that discussed in Section 3.3.8 for monopoles. The kink U(1) charge is no longer integer in the presence of the θ term, it is shifted by $\theta/(2\pi)$.

3.5.3 The kink mass and holomorphy

Taking account of $E_{[\alpha]}$ – the energy of an "internal motion" – the kink mass can be written as

$$
\begin{aligned}
M &= \frac{2|m|}{g^2} + \frac{g^2|m|}{4}\left(k + \frac{\theta}{2\pi}\right)^2 \\
&= \frac{2|m|}{g^2}\left\{1 + \frac{g^4}{4}\left(k + \frac{\theta}{2\pi}\right)^2\right\}^{1/2} \\
&\doteq 2|m|\left|\frac{1}{g^2} + i\,\frac{\theta + 2\pi k}{4\pi}\right|.
\end{aligned}
\tag{3.5.20}
$$

Formally, the second equality here is approximate, valid to the leading order in the coupling constant. In fact, it is exact. The important circumstance to be stressed is that the kink mass depends on a special combination of the coupling constant and θ, namely,

$$
\tau = \frac{1}{g^2} + i\,\frac{\theta}{4\pi}
\tag{3.5.21}
$$

In other words, it is the complexified coupling constant that enters.

Note that g^2 in Eq. (3.5.20) is the bare coupling constant. It is quite clear that the kink mass, being a physical parameter, should contain the renormalized constant $g^2(m)$, after taking account of radiative corrections.

Since the kink mass $M = |Z|$ radiative corrections must replace the bare $1/g^2$ by the renormalized $1/g^2(m)$ in Z. One-loop calculation is quite trivial. First, rotate the mass parameter m in such a way as to make it real, $m \to |m|$. Simultaneously, the θ angle is replaced by an effective θ,

$$
\theta \to \theta_{\text{eff}} = \theta + 2\beta,
\tag{3.5.22}
$$

where the phase β is defined in Eq. (3.5.11). Next, decompose the field ϕ into a classical plus quantum part,

$$
\phi \to \phi + \delta\phi.
$$

Then the D part of the central charge

$$
Z = mq - i\int dz\,\partial_z\,m\,D
$$

becomes

$$
D \to D + \frac{2}{g^2}\,\frac{1 - \phi^\dagger\phi}{\left(1 + \bar\phi\phi\right)^3}\,\delta\phi^\dagger\,\delta\phi.
\tag{3.5.23}
$$

(The term in parentheses in (3.5.6) – the anomaly – gives a non-logarithmic contribution which we ignore for the time being.) Contracting $\delta\phi^\dagger \, \delta\phi$ into a loop and calculating this loop we arrive at

$$D \rightarrow \frac{\phi^\dagger \phi}{\chi} \left[\frac{2}{g^2} - \frac{2}{4\pi} \ln \frac{M_{uv}^2}{|m|^2} \right], \qquad (3.5.24)$$

which, in turn, yields

$$Z = 2im \left\{ \tau - \frac{1}{4\pi} \ln \frac{M_{uv}^2}{m^2} - i \frac{k}{2} \right\} \equiv 2im \left\{ \tau_{ren} - i \frac{k}{2} \right\}. \qquad (3.5.25)$$

A salient feature of this formula is the holomorphic dependence of Z on m and τ. Such holomorphic dependence would be impossible if two and more loops contributed to D renormalization. Thus, D renormalization beyond one loop must cancel, and it does.[13] Note also that the bare coupling in Eq. (3.5.25) conspires with the logarithm in such a way as to replace the bare coupling by that renormalized at $|m|$, as was expected.

The analysis carried out above is quasiclassical. It tells us nothing about possible occurrence of nonperturbative terms in Z. In fact, all terms of the type

$$\left\{ \frac{M_{uv}^2}{m^2} \exp\left(-4\pi\tau\right) \right\}^\ell, \quad \ell = \text{integer}$$

are fully compatible with holomorphy; they can and do emerge from instantons. An indirect calculation of nonperturbative terms was performed in Ref. [30].

The exact formula for this central charge obtained by Dorey is

$$Z = mq - im_D T, \qquad (3.5.26)$$

where the subscript D in m_D appears for historical reasons, in parallel with the Seiberg–Witten solution (it stands for dual), and

$$m_D = \frac{m}{\pi} \left[\frac{1}{2} \ln \frac{m + \sqrt{m^2 + 4\Lambda^2}}{m - \sqrt{m^2 + 4\Lambda^2}} - \sqrt{1 + \frac{4\Lambda^2}{m^2}} \right]. \qquad (3.5.27)$$

Furthermore, T is the topological charge of the kink under consideration, $T = \pm 1$. The limit $|m|/\Lambda \rightarrow \infty$ corresponds to the quasiclassical domain, while corrections of the type $(\Lambda/m)^{2k}$ are induced by instantons.

---

[13] Fermions are important for this cancellation.

It is instructive to consider the quasiclassical limit of Eq. (3.5.27) when the mass m is real and large, $m \gg \Lambda$. In this limit

$$\langle Z \rangle_{\text{kink}} = -\frac{i\,m}{2\pi}\left[\ln\left(-\frac{m^2}{\Lambda^2}\right) - 2\right]$$

$$= \frac{1}{2}m - im\left(\frac{2}{g_0^2} - \frac{1}{2\pi}\ln\frac{M_{\text{uv}}^2}{m^2}\right) + \frac{i\,m}{\pi}, \qquad (3.5.28)$$

where g_0^2 is the bare coupling constant, and M_{uv} is the ultraviolet cut off. The first term in the second line reflects the fractional U(1) charge, $q = 1/2$, carried by the soliton at $\theta = 0$. The reason for the occurrence of half-integer charge will be explained in detail in Section 3.5.5. The second term coincides with the one-loop corrected average of $(-i \int dz\, \partial_z O_{\text{canon}})$ in the central charge. The third term im/π represents the anomaly.

What happens when one travels from the domain of large $|m|$ to that of small $|m|$? If $m = 0$ we know (e.g. from the mirror representation [120]) that there are two degenerate two-dimensional kink supermultiplets, corresponding to the Cecotti–Fendley–Intriligator–Vafa (CFIV) index $= 2$ [121]. They have quantum numbers $\{q, T\} = (0, 1)$ and $(1, 1)$, respectively. Away from the point $m = 0$ the masses of these states are no longer equal; there is one singular point with one of the two states becoming massless [34]. The region containing the point $m = 0$ is separated from the quasiclassical region of large m by the curve of marginal stability (CMS) on which an infinite number of other BPS states, visible quasiclassically, decay, see Fig. 3.13.[14] Thus, the infinite tower of the $\{q, T\}$ BPS states existing in the quasiclassical domain degenerates in just two stable BPS states in the vicinity of $m = 0$.

3.5.4 Fermions in quasiclassical consideration

Non-zero modes are irrelevant for our consideration since, being combined with the boson non-zero modes, they cancel for critical solitons, a usual story. Thus, for our purposes it is sufficient to focus on the (static) zero modes in the kink background (3.5.15). The coefficients in front of the fermion zero modes will become (time-dependent) fermion moduli, for which we are going to build corresponding quantum mechanics. There are two such moduli, $\bar{\eta}$ and η.

---

[14] CMS for CP($N - 1$) with $N > 2$ is considered in [122].

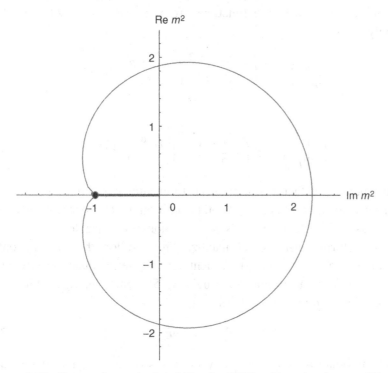

Figure 3.13. Curve of marginal stability in CP(1) with twisted mass. We set $4\Lambda^2 \rightarrow 1$. From Ref. [34].

The equations for the fermion zero modes are

$$\partial_z \Psi_L - \frac{2}{\chi} \left(\bar{\phi}\partial_z\phi\right) \Psi_L - i \frac{1 - \bar{\phi}\phi}{\chi} |m| e^{i\beta} \Psi_R = 0,$$

$$\partial_z \Psi_R - \frac{2}{\chi} \left(\bar{\phi}\partial_z\phi\right) \Psi_R + i \frac{1 - \bar{\phi}\phi}{\chi} |m| e^{-i\beta} \Psi_L = 0 \qquad (3.5.29)$$

(plus similar equations for $\bar{\Psi}$; since our operator is Hermitean we do not need to consider them separately).

It is not difficult to find a solution to these equations, either directly, or using supersymmetry. Indeed, if we know the bosonic solution (3.5.15), its fermionic superpartner – and the fermion zero modes are such superpartners – is obtained from the bosonic one by those two supertransformations which act on $\bar{\phi}$, ϕ nontrivially. In this way we conclude that the functional form of the fermion zero mode must coincide with the functional form of the boson solution (3.5.15). Concretely,

$$\begin{pmatrix} \Psi_R \\ \Psi_L \end{pmatrix} = \eta \left(\frac{g^2 |m|}{2} \right)^{1/2} \begin{pmatrix} -ie^{-i\beta} \\ 1 \end{pmatrix} e^{|m|(z-z_0)} \qquad (3.5.30)$$

and

$$\begin{pmatrix} \bar{\Psi}_R \\ \bar{\Psi}_L \end{pmatrix} = \bar{\eta} \left(\frac{g^2 |m|}{2} \right)^{1/2} \begin{pmatrix} ie^{i\beta} \\ 1 \end{pmatrix} e^{|m|(z-z_0)}, \qquad (3.5.31)$$

where the numerical factor is introduced to ensure proper normalization of the quantum-mechanical Lagrangian. Another solution which asymptotically, at large z, behaves as $e^{3|m|(z-z_0)}$ must be discarded as non-normalizable.

Now, to perform quasiclassical quantization we follow the standard route: the moduli are assumed to be time-dependent, and we derive quantum mechanics of moduli starting from the original Lagrangian (3.5.1). Substituting the kink solution and the fermion zero modes for Ψ one gets

$$\mathcal{L}'_{QM} = i\,\bar{\eta}\dot{\eta}. \qquad (3.5.32)$$

In the Hamiltonian approach the only remnants of the fermion moduli are the anticommutation relations

$$\{\bar{\eta}\eta\} = 1, \quad \{\bar{\eta}\bar{\eta}\} = 0, \quad \{\eta\eta\} = 0, \qquad (3.5.33)$$

which tell us that the wave function is two-component (i.e. the kink supermultiplet is two-dimensional). One can implement Eq. (3.5.33) by choosing e.g. $\bar{\eta} = \sigma^+$, $\eta = \sigma^-$.

The fact that there are two critical kink states in the supermultiplet is consistent with the multiplet shortening in $\mathcal{N} = 2$. Indeed, in two dimensions the full $\mathcal{N} = 2$ supermultiplet must consist of four states: two bosonic and two fermionic. 1/2 BPS multiplets are shortened – they contain twice less states than the full supermultiplets, one bosonic and one fermionic. This is to be contrasted with the single-state kink supermultiplet in the minimal supersymmetric model of Section 3.1.1. The notion of the fermion parity remains well-defined in the kink sector of the CP(1) model.

3.5.5 Combining bosonic and fermionic moduli

Quantum dynamics of the kink at hand is summarized by the Hamiltonian

$$H_{QM} = \frac{M_0}{2} \dot{\bar{\zeta}}\dot{\zeta} \qquad (3.5.34)$$

acting in the space of two-component wave functions. The variable ζ here is a complexified kink center,

$$\zeta = z_0 + \frac{i}{|m|} \alpha. \qquad (3.5.35)$$

For simplicity, we set the vacuum angle $\theta = 0$ for the time being (it will be reinstated later).

The original field theory we deal with has four conserved supercharges. Two of them, $\mathcal{Q}$ and $\bar{\mathcal{Q}}$, see Eq. (3.5.11), act trivially in the critical kink sector. In moduli quantum mechanics they take the form

$$\mathcal{Q} = \sqrt{M_0}\, \dot{\zeta} \eta, \quad \bar{\mathcal{Q}} = \sqrt{M_0}\, \dot{\bar{\zeta}} \bar{\eta}; \qquad (3.5.36)$$

they do indeed vanish provided that the kink is at rest. Superalgebra describing kink quantum mechanics is $\{\bar{\mathcal{Q}}\,\mathcal{Q}\} = 2H_{\text{QM}}$. This is nothing but Witten's $\mathcal{N} = 1$ supersymmetric quantum mechanics [123] (two supercharges). The realization we deal with is peculiar and distinct from that of Witten. Indeed, the standard Witten quantum mechanics includes one (real) bosonic degree of freedom and two fermionic, while we have two bosonic degrees of freedom, x_0 and α. Nevertheless, superalgebra remains the same due to the fact that the bosonic coordinate is complexified.

Finally, to conclude this section, let us calculate the U(1) charge of the kink states. We start from Eq. (3.5.8), substitute the fermion zero modes and get[15]

$$\Delta q_{\text{U}(1)} = \frac{1}{2}[\bar{\eta}\eta] \qquad (3.5.37)$$

(this is to be added to the bosonic part, Eq. (3.5.19)). Given that $\bar{\eta} = \sigma^+$ and $\eta = \sigma^-$ we arrive at $\Delta q_{\text{U}(1)} = \frac{1}{2}\sigma_3$. This means that the U(1) charges of two kink states in the supermultiplet split from the value given in Eq. (3.5.19): one has the U(1) charge

$$k + \frac{1}{2} + \frac{\theta}{2\pi},$$

---

[15] To set the scale properly, so that the U(1) charge of the vacuum state vanishes, one must antisymmetrize the fermion current, $\bar{\Psi}\gamma^\mu\Psi \to (1/2)\left(\bar{\Psi}\gamma^\mu\Psi - \bar{\Psi}^c\gamma^\mu\Psi^c\right)$ where the superscript c denotes C conjugation.

and another

$$k - \frac{1}{2} + \frac{\theta}{2\pi}.$$

In this way we explain the occurrence of $1/2$ seen from the quasiclassical expansion of the exact formula (3.5.28).

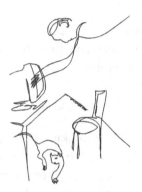

Part II

Long journey

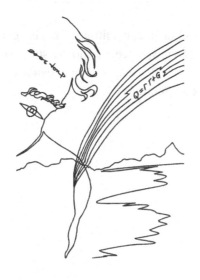

Warning

In Part II we switch to Euclidean conventions. This is appropriate and convenient from the technical point of view, since we mostly study static (time-independent) field configurations, and $A_0 = 0$. Then the Euclidean action reduces to the energy functional. See Appendix A, Section A.7.

Introduction to Part II

Ever since 't Hooft [124] and Mandelstam [125] put forward the hypothesis of the dual Meissner effect to explain color confinement in non-Abelian gauge theories, people were trying to find a controllable approximation in which one could reliably demonstrate the occurrence of the dual Meissner effect in these theories. A breakthrough achievement was the Seiberg–Witten solution [2] of $\mathcal{N} = 2$ supersymmetric Yang–Mills theory. They found massless monopoles and, adding a small ($\mathcal{N} = 2$)-breaking deformation, proved that they condense creating strings carrying a chromoelectric flux. It was a great success in qualitative understanding of color confinement.

A more careful examination shows, however, that details of the Seiberg–Witten confinement are quite different from those we expect in QCD-like theories. Indeed, a crucial aspect of Ref. [2] is that the SU(N) gauge symmetry is first broken, at a high scale, down to U(1)$^{N-1}$, which is then completely broken at a much lower scale where condensation of magnetic monopoles occurs. Correspondingly, the strings in the Seiberg–Witten solution are, in fact, Abelian strings [36] of the Abrikosov–Nielsen–Olesen (ANO) type which results, in turn, in confinement whose structure does not resemble at all that of QCD.[1] In particular, the "hadronic" spectrum is much richer than that in QCD [126, 127, 128, 35, 129]. To see this it is sufficient to observe that, given the low-energy gauge group U(1)$^{N-1}$, one has $N - 1$ Abelian strings associated with each of the $N - 1$ Abelian factors. Since

$$\pi_1\big(U(1)^{N-1}\big) = Z^{N-1},\qquad (3.5.38)$$

---

[1] It is believed, however, that the transition from this Abelian to non-Abelian confinement is smooth. The Seiberg–Witten construction has an adjustable parameter μ which governs the strength of the ($\mathcal{N} = 2$)-breaking deformation. In passing from the Abelian strings at small μ to non-Abelian at large μ no phase transition is expected. One cannot *prove* this statement at present because the large-μ side is inaccessible. In contrast, in the models to be discussed below, we can address and explore *directly* non-Abelian strings and associated dynamics.

the Abelian strings and, therefore, the meson spectrum come in $N-1$ infinite towers. This feature is not expected in real-world QCD. Moreover, there is no experimental indication of dynamical Abelization in QCD.

Here in Part II we begin our long journey which covers the advances of the last decade. Most developments are even fresher, they refer to the last five years or so. First we dwell on the recent discovery of non-Abelian strings [130, 131, 132, 133] which appear in certain regimes in $\mathcal{N} = 2$ supersymmetric gauge theories. Moreover, they were found even in $\mathcal{N} = 1$ theories, the so-called M model, see Section 5.2. The most important feature of these strings is that they acquire orientational zero modes associated with rotations of their color flux inside a non-Abelian $SU(N)$ subgroup of the gauge group. The occurrence of these zero modes makes these strings non-Abelian.

The flux tubes in non-Abelian theories at weak coupling were studied in the past in numerous papers [134, 135, 136, 137, 138, 139, 140]. These strings are referred to as Z_N strings because they are related to the center of the gauge group $SU(N)$. Consider, say, the $SU(N)$ gauge theory with a few (more than one) scalar fields in the adjoint representation. Suppose the adjoint scalars condense in such a way that the $SU(N)$ gauge group is broken down to its center Z_N. Then string solutions are classified according to

$$\pi_1\left(\frac{SU(N)}{Z_N}\right) = Z_N.$$

In all these previous constructions [134, 135, 136, 137, 138, 139, 140] of the Z_N strings the flux was always directed in a fixed group direction (corresponding to a Cartan subalgebra), and no moduli that would allow to freely rotate its orientation in the group space were ever obtained. Therefore it is reasonable to call these Z_N strings Abelian, in contrast with the non-Abelian strings, to be discussed below, which have orientational moduli.

Consideration of non-Abelian strings naturally leads us to confined non-Abelian monopoles. We follow the fate of the classical 't Hooft–Polyakov monopole (classical not in the sense of "non-quantum" but rather in the sense of something belonging to textbooks) in the Higgs "medium" – from free monopoles, through a weakly confined regime, to a highly quantum regime in which confined monopoles manifest themselves as kinks in the low-energy theory on the string world sheet. The confined monopoles are sources (sinks) to which the magnetic flux tubes are attached. We demonstrate that they are dual to quarks just in the same vein as the magnetic flux tubes are dual to the electric ones.

Our treatise covers the non-Abelian flux tubes both in theories with the minimal ($\mathcal{N} = 1$) and extended ($\mathcal{N} = 2$) supersymmetry. The world sheet theory for the Abelian strings contains only translational and supertranslational moduli fields.

At the same time, for the non-Abelian strings the world sheet theory acquires additional massless (or very light) fields. Correspondingly, the Lüscher term coefficient, which counts the number of such degrees of freedom, changes [141].

The third element of the big picture which we explore is the wall-string junction. From string/D brane theory it is well known that fundamental strings can end on the brane. In fact, this is a defining property of the brane. Since our task is to reveal in gauge theories all phenomena described by string/D brane theory we must be able to see the string-wall junctions. And we do see them! The string-wall junctions which later got the name *boojums* were first observed in an $\mathcal{N} = 2$ gauge theory in Ref. [142]. This construction as well as later advances in the boojum theory are thoroughly discussed in Part II.

The domain walls as brane prototypes must possess another remarkable feature – they must localize gauge fields. This localization was first proven to occur in an $\mathcal{N} = 2$ gauge theory in Ref. [37]. The domain-wall world sheet theory is the theory of three-dimensional gauge fields after all!

All the above elements combined together lead us to a thorough understanding of the Meissner effect in non-Abelian theories. To understand QCD we need to develop a model of the *dual* Meissner effect. Although this problem is not yet fully solved, we report here on significant progress in this direction.

4

Non-Abelian strings

In this chapter we discuss a particular class of $\mathcal{N} = 2$ supersymmetric gauge theories in which non-Abelian strings were found. One can pose the question: what is so special about these models that makes an Abelian Z_N string become non-Abelian? Models we will dwell on below have both gauge and flavor symmetries broken by the condensation of scalar fields. The common feature of these models is that some global diagonal combination of color and flavor groups survive the breaking. We consider the case when this diagonal group is $SU(N)_{C+F}$, where the subscript $C + F$ means a combination of global color and flavor groups. The presence of this unbroken subgroup is responsible for the occurrence of the orientational zero modes of the string which entail its non-Abelian nature.

Clearly, the presence of supersymmetry is not important for the construction of non-Abelian strings. In particular, while here we focus on the BPS non-Abelian strings in $\mathcal{N} = 2$ supersymmetric gauge theories, in Chapter 5 we review non-Abelian strings in $\mathcal{N} = 1$ supersymmetric theories and in Chapter 6 in non-supersymmetric theories.

4.1 Basic model: $\mathcal{N} = 2$ SQCD

The model we will deal with derives from $\mathcal{N} = 2$ SQCD with the gauge group $SU(N + 1)$ and $N_f = N$ flavors of the fundamental matter hypermultiplets which we will call quarks [3]. At a generic point on the Coulomb branch of this theory, the gauge group is broken down to $U(1)^N$. We will be interested, however, in a particular subspace of the Coulomb branch, on which the gauge group is broken down to $SU(N) \times U(1)$. We will enforce this regime by a special choice of the quark mass terms.

The breaking $SU(N + 1) \rightarrow SU(N) \times U(1)$ occurs at the scale m which is supposed to lie very high, $m \gg \Lambda_{SU(N+1)}$, where $\Lambda_{SU(N+1)}$ is the scale of the $SU(N + 1)$ theory. Correspondingly, the masses of the gauge bosons from the

SU$(N + 1)/$SU$(N)\times$U(1) sector and their superpartners, are very large – proportional to m – and so are the masses of the $(N + 1)$-th color component of the quark fields in the fundamental representation. We will be interested in the phenomena at the scales $\ll m$. Therefore, our starting point is in fact the SU$(N)\times$U(1) model with $N_f = N$ matter fields in the fundamental representation of SU(N), as it emerges after the SU$(N+1) \to$ SU$(N)\times$U(1) breaking. These matter fields are also coupled to the U(1) gauge field.

The field content of SU$(N)\times$U(1) $\mathcal{N} = 2$ SQCD with N flavors is as follows. The $\mathcal{N} = 2$ vector multiplet consists of the U(1) gauge field A_μ and the SU(N) gauge field A_μ^a, (here $a = 1, \ldots, N^2 - 1$), and their Weyl fermion superpartners $(\lambda_\alpha^1, \lambda_\alpha^2)$ and $(\lambda_\alpha^{1a}, \lambda_\alpha^{2a})$, plus complex scalar fields a, and a^a. The latter are in the adjoint representation of SU(N). The spinorial index of λ's runs over $\alpha = 1, 2$. In this sector the global SU$(2)_R$ symmetry inherent to the $\mathcal{N} = 2$ model at hand manifests itself through rotations $\lambda^1 \leftrightarrow \lambda^2$.

The quark multiplets of the SU$(N)\times$U(1) theory consist of the complex scalar fields q^{kA} and $\tilde{q}_{Ak}$ (squarks) and the Weyl fermions ψ^{kA} and $\tilde{\psi}_{Ak}$, all in the fundamental representation of the SU(N) gauge group. Here $k = 1, \ldots, N$ is the color index while A is the flavor index, $A = 1, \ldots, N$. Note that the scalars q^{kA} and $\bar{\tilde{q}}^{kA}$ form a doublet under the action of the global SU$(2)_R$ group.

Then the original SU$(N + 1)$ theory is perturbed by adding a small mass term for the adjoint matter, via the superpotential $\mathcal{W} = \mu \, \text{Tr} \, \Phi^2$. Generally speaking, this superpotential breaks $\mathcal{N} = 2$ down to $\mathcal{N} = 1$. The Coulomb branch shrinks to a number of isolated $\mathcal{N} = 1$ vacua [2, 3, 126, 143, 144]. In the limit of $\mu \to 0$ these vacua correspond to special singular points on the Coulomb branch in which N monopoles/dyons or quarks become massless. The first $(N + 1)$ of these points (often referred to as the Seiberg–Witten vacua) are always at strong coupling. They correspond to $\mathcal{N} = 1$ vacua of the pure SU$(N + 1)$ gauge theory.

The massless quark points – they present vacua of a distinct type, to be referred to as the quark vacua – may or may not be at weak coupling depending on the values of the quark mass parameters m_A. If $m_A \gg \Lambda_{\text{SU}(N+1)}$, the quark vacua do lie at weak coupling. Below we will be interested only in these quark vacua assuming that the condition $m_A \gg \Lambda_{\text{SU}(N+1)}$ is met.

In the low-energy SU$(N)\times$U(1) theory, which is our starting point, the perturbation $\mathcal{W} = \mu \, \text{Tr} \, \Phi^2$ can be truncated, leading to a crucial simplification. Indeed, since the $\mathcal{A}$ chiral superfield, the $\mathcal{N} = 2$ superpartner of the U(1) gauge field,[1]

$$\mathcal{A} \equiv a + \sqrt{2}\lambda^2\theta + F_a \theta^2, \tag{4.1.1}$$

---

[1] The superscript 2 in Eq. (4.1.1) is the global SU$(2)_R$ index of λ rather than λ squared.

it not charged under the gauge group SU(N)×U(1), one can introduce a superpotential linear in $\mathcal{A}$,

$$\mathcal{W}_{\mathcal{A}} = -\frac{N}{2\sqrt{2}}\, \xi\, \mathcal{A}. \tag{4.1.2}$$

Here we expand Tr Φ^2 around its vacuum expectation value (VEV), and truncate
the series keeping only the linear term in $\mathcal{A}$. The truncated superpotential is a
Fayet–Iliopoulos (FI) F-term.

Let us explain this in more detail. In $\mathcal{N} = 1$ supersymmetric theory with the
gauge group SU(N)×U(1) one can add the following FI term to the action [145]
(we will call it the FI D-term here):

$$\xi_3\, D \tag{4.1.3}$$

where D is the D-component of the U(1) gauge superfield. In $\mathcal{N} = 2$ SUSY theory
the field D belongs to the SU(2)$_R$ triplet, together with the F components of the
chiral field $\mathcal{A}$, (F and $\bar{F}$). Namely, let us introduce a triplet F_p ($p = 1, 2, 3$) using
the relations [2]

$$D = F_3,$$
$$F_{\mathcal{A}} = \frac{1}{\sqrt{2}}\,(F_1 + i F_2),$$
$$\bar{F}_{\mathcal{A}} = \frac{1}{\sqrt{2}}\,(F_1 - i F_2). \tag{4.1.4}$$

Now, the generalized FI term can be written as

$$S_{\text{FI}} = -\frac{N}{2} \int d^4x \sum_p \xi_p F_p. \tag{4.1.5}$$

Comparing this with Eq. (4.1.2) we identify

$$\xi = (\xi_1 - i\xi_2),$$
$$\bar{\xi} = (\xi_1 + i\xi_2) \tag{4.1.6}$$

This is the reason why we refer to the superpotential (4.1.2) as to the FI F-term.

A remarkable feature of the FI term is that it does *not* break $\mathcal{N} = 2$ supersymmetry [127, 35]. Keeping higher order terms of the expansion of μ Tr Φ^2 in
powers of $\mathcal{A}$ would inevitably explicitly break $\mathcal{N} = 2$. For our purposes it is crucial

---

[2] Attention: The index p is an SU(2)$_R$ index rather than the color index!

that the model we will deal with is *exactly* $\mathcal{N} = 2$ supersymmetric. This ensures that the flux tube solutions of the model are BPS-saturated. If higher order terms in $\mathcal{A}$ are taken into account, $\mathcal{N} = 2$ supersymmetry is broken down to $\mathcal{N} = 1$ and strings are no longer BPS, generally speaking. The superconductivity in the model becomes of type I [35].

4.1.1 SU(N)×U(1) $\mathcal{N} = 2$ QCD

The bosonic part of our SU(N)×U(1) theory has the form [131]

$$
S = \int d^4x \left[\frac{1}{4g_2^2}(F_{\mu\nu}^a)^2 + \frac{1}{4g_1^2}(F_{\mu\nu})^2 + \frac{1}{g_2^2}|D_\mu a^a|^2 + \frac{1}{g_1^2}|\partial_\mu a|^2 \right.
$$
$$
\left. + |\nabla_\mu q^A|^2 + |\nabla_\mu \bar{\tilde{q}}^A|^2 + V(q^A, \tilde{q}_A, a^a, a) \right]. \tag{4.1.7}
$$

Here D_μ is the covariant derivative in the adjoint representation of SU(N), and

$$
\nabla_\mu = \partial_\mu - \frac{i}{2} A_\mu - i A_\mu^a T^a. \tag{4.1.8}
$$

We suppress the color SU(N) indices, and T^a are the SU(N) generators normalized as

$$
\text{Tr}\,(T^a T^b) = (1/2)\,\delta^{ab}.
$$

The coupling constants g_1 and g_2 correspond to the U(1) and SU(N) sectors, respectively. With our conventions, the U(1) charges of the fundamental matter fields are $\pm 1/2$.

The potential $V(q^A, \tilde{q}_A, a^a, a)$ in the action (4.1.7) is a sum of D and F terms,

$$
V(q^A, \tilde{q}_A, a^a, a) = \frac{g_2^2}{2} \left(\frac{i}{g_2^2} f^{abc} \bar{a}^b a^c + \bar{q}_A T^a q^A - \tilde{q}_A T^a \bar{\tilde{q}}^A \right)^2
$$
$$
+ \frac{g_1^2}{8} \left(\bar{q}_A q^A - \tilde{q}_A \bar{\tilde{q}}^A - N\xi_3 \right)^2
$$
$$
+ 2g_2^2 |\tilde{q}_A T^a q^A|^2 + \frac{g_1^2}{2} \left| \tilde{q}_A q^A - \frac{N}{2}\xi \right|^2
$$
$$
+ \frac{1}{2} \sum_{A=1}^{N} \left\{ |(a + \sqrt{2}m_A + 2T^a a^a)q^A|^2 \right.
$$
$$
\left. + |(a + \sqrt{2}m_A + 2T^a a^a)\bar{\tilde{q}}^A|^2 \right\}. \tag{4.1.9}
$$

Here f^{abc} stand for the structure constants of the SU(N) group, and the sum over the repeated flavor indices A is implied.

The first and second lines represent D terms, the third line the F_A terms, while the fourth and the fifth lines represent the squark F terms. Using the SU(2)$_R$ rotations we can always direct the FI parameter vector ξ_p in a given direction. Below in most cases we will align the FI F-term to make the parameter ξ real. In other words,

$$\xi_3 = 0, \quad \xi_2 = 0, \quad \xi = \xi_1. \tag{4.1.10}$$

4.1.2 The vacuum structure and excitation spectrum

Now we briefly review the vacuum structure and the excitation spectrum in our basic SU(N)$\times$U(1) model. As was mentioned, the underlying $\mathcal{N} = 2$ SQCD with the gauge group SU($N + 1$) has a variety of vacua [143, 144, 140]. In addition to N strong coupling vacua which exist in pure gauge theory, there is a number of the so-called r quark vacua, where r is the number of the quark flavors which develop VEV's in the given vacuum. We will limit ourselves[3] to a particular isolated vacuum, with the maximal possible value of r,

$$r = N.$$

The vacua of the theory (4.1.7) are determined by the zeros of the potential (4.1.9). The adjoint fields develop the following VEV's:

$$\langle \Phi \rangle = -\frac{1}{\sqrt{2}} \begin{pmatrix} m_1 & \cdots & 0 \\ \cdots & \cdots & \cdots \\ 0 & \cdots & m_N \end{pmatrix}, \tag{4.1.11}$$

where we defined the scalar adjoint matrix as

$$\Phi = \frac{1}{2} a + T^a a^a. \tag{4.1.12}$$

For generic values of the quark masses, the SU(N) subgroup of the gauge group is broken down to U(1)$^{N-1}$. However, for a special choice

$$m_1 = m_2 = \cdots = m_N, \tag{4.1.13}$$

---

[3] There are singular points on the Coulomb branch of the underlying SU($N + 1$) theory where more than N quark flavors become massless. These singularities are the roots of Higgs branches [143, 144, 140].

which we will be mostly interested in in this section, the $SU(N) \times U(1)$ gauge group remains classically unbroken. In fact, the common value m of the quark masses determines the scale of breaking of the $SU(N + 1)$ gauge symmetry of the underlying theory down to $SU(N) \times U(1)$ gauge symmetry of our benchmark low-energy theory (4.1.7).

If the value of the FI parameter is taken real we can exploit gauge rotations to make the quark VEV's real too. Then in the case at hand they take the color-flavor locked form

$$\langle q^{kA} \rangle = \langle \bar{\bar{q}}^{kA} \rangle = \sqrt{\frac{\xi}{2}} \begin{pmatrix} 1 & \cdots & 0 \\ \cdots & \cdots & \cdots \\ 0 & \cdots & 1 \end{pmatrix},$$
$$k = 1, \ldots, N, \quad A = 1, \ldots, N, \tag{4.1.14}$$

where we write down the quark fields as an $N \times N$ matrix in the color and flavor indices. This particular form of the squark condensates is dictated by the third line in Eq. (4.1.9). Note that the squark fields stabilize at non-vanishing values entirely due to the U(1) factor represented by the second term in the third line.

The vacuum field (4.1.14) results in the spontaneous breaking of both gauge and flavor $SU(N)$'s. A diagonal global $SU(N)$ survives, however,

$$U(N)_{\text{gauge}} \times SU(N)_{\text{flavor}} \to SU(N)_{C+F}. \tag{4.1.15}$$

Thus, a color-flavor locking takes place in the vacuum. A version of this pattern of the symmetry breaking was suggested long ago [146].

Let us move on to the issue of the excitation spectrum in this vacuum [35, 131]. The mass matrix for the gauge fields (A_μ^a, A_μ) can be read off from the quark kinetic terms in Eq. (4.1.7). It shows that all $SU(N)$ gauge bosons become massive, with one and the same mass

$$M_{\text{SU}(N)} = g_2 \sqrt{\xi}. \tag{4.1.16}$$

The equality of the masses is no accident. It is a consequence of the unbroken $SU(N)_{C+F}$ symmetry (4.1.15).

The mass of the U(1) gauge boson is

$$M_{\text{U}(1)} = g_1 \sqrt{\frac{N}{2} \xi}. \tag{4.1.17}$$

Thus, the theory is fully Higgsed. The mass spectrum of the adjoint scalar excitations is the same as the one for the gauge bosons. This is enforced by $\mathcal{N} = 2$.

What is the mass spectrum of the quark excitations? It can be read off from the potential (4.1.9). We have $4N^2$ real degrees of freedom of quark scalars q and $\tilde{q}$. Out of those N^2 are eaten up by the Higgs mechanism. The remaining $3N^2$ states split in three plus $3(N^2 - 1)$ states with masses (4.1.17) and (4.1.16), respectively. Combining these states with the massive gauge bosons and the adjoint scalar states we get [35, 131] one long $\mathcal{N} = 2$ BPS multiplet (eight real bosonic plus eight fermionic degrees of freedom) with mass (4.1.17) and $N^2 - 1$ long $\mathcal{N} = 2$ BPS multiplets with mass (4.1.16). Note that these supermultiplets come in representations of the unbroken SU(N)$_{C+F}$ group, namely, the singlet and the adjoint representations.

To conclude this section we want to discuss quantum effects in the theory (4.1.7). At a high scale m the SU($N + 1$) gauge group is broken down to SU(N)$\times$U(1) by condensation of the adjoint fields if the condition (4.1.13) is met. The SU(N) sector is asymptotically free. The running of the corresponding gauge coupling, if non-interrupted, would drag the theory into the strong coupling regime. This would invalidate our quasiclassical analysis. Moreover, strong coupling effects on the Coulomb branch would break SU(N) gauge subgroup (as well as the SU(N)$_{C+F}$ group) down to U(1)$^{N-1}$ by the Seiberg–Witten mechanism [2]. No non-Abelian strings would emerge.

A possible way out was proposed in [143, 144]. One can add more flavors to the theory making $N_f > 2N$. Then the SU(N) sector is not asymptotically free and does not evolve into the strong coupling regime. However, the ANO strings in the multiflavor theory (on the Higgs branches) become semilocal strings [147] and confinement is lost (see Section 4.7).

Here we take a different route assuming the FI parameter ξ to be large,[4]

$$\xi \gg \Lambda_{\text{SU}(N)}. \tag{4.1.18}$$

This condition ensures weak coupling in the SU(N) sector because the SU(N) gauge coupling does not run below the scale of the quark VEV's which is determined by ξ. More explicitly,

$$\frac{8\pi^2}{g_2^2(\xi)} = N \ln \frac{\sqrt{\xi}}{\Lambda_{\text{SU}(N)}} \gg 1. \tag{4.1.19}$$

---

[4] We discuss this important issue in more detail at the end of Section 4.9.

Alternatively one can say that

$$\Lambda_{SU(N)}^N = \xi^{N/2} \exp\left(-\frac{8\pi^2}{g_2^2(\xi)}\right) \ll \xi^{N/2}. \qquad (4.1.20)$$

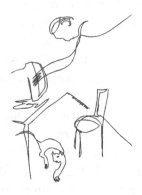

4.2 Z_N Abelian strings

Strictly speaking, $\mathcal{N} = 2$ SQCD with the gauge group SU($N + 1$) does not have stable flux tubes. They are unstable due to monopole–antimonopole pair creation in the SU($N + 1$)/SU(N)×U(1) sector. However, at large m these monopoles become heavy. In fact, there are no such monopoles in the low-energy theory (4.1.7) (where they can be considered as infinitely heavy). Therefore, the theory (4.1.7) has stable string solutions. When the perturbation $\mu \, \mathrm{Tr} \, \Phi^2$ is truncated to the FI term (4.1.2), the theory enjoys $\mathcal{N} = 2$ supersymmetry and has BPS string solutions [127, 35, 148, 140, 131]. Note that here we discuss magnetic flux tubes. They are formed in the Higgs phase of the theory upon condensation of the squark fields and lead to confinement of monopoles.

Now, let us briefly review the BPS string solutions [140, 130, 131] in the model (4.1.7). We will consider the case of equal quark mass terms (4.1.13) when the global SU(N)$_{C+F}$ group is unbroken. First we review the Abelian solutions for Z_N strings and then, in Section 4.3 show that in the limit $m_1 = m_2 = \cdots = m_N \equiv m$ they acquire orientational moduli.

In fact, the Z_N Abelian strings considered below are just partial solutions of the vortex equations (see Eq. (4.2.10) below). In the equal mass limit (4.1.13) the global SU(N)$_{C+F}$ group is restored and the general solution for the non-Abelian string gets a continuous moduli space isomorphic to CP($N - 1$). The Z_N strings are just N discrete points on this moduli space.

In the generic case of unequal quark masses, the $SU(N)_{C+F}$ group is explicitly broken, and the continuous moduli space of the string solutions is lifted. Only the Z_N Abelian strings survive this breaking. We will dwell on the case of generic quark masses in Section 4.4.4.

It turns out that the string solutions do not involve the adjoint fields a and a^a. The BPS strings are "built" from gauge and quark fields only. Therefore, in order to find the classical solution, in the action (4.1.7) we can set the adjoint fields to their VEV's (4.1.11). This is consistent with equations of motion. Of course, at the quantum level the adjoint fields start fluctuating, deviating from their VEV's.

We use the *ansatz*

$$q^{kA} = \bar{\tilde{q}}^{kA} = \frac{1}{\sqrt{2}}\varphi^{kA} \qquad (4.2.1)$$

reducing the number of the squark degrees of freedom to one complex field for each color and flavor. With these simplifications the action of the model (4.1.7) becomes

$$S = \int d^4x \left\{ \frac{1}{4g_2^2}(F_{\mu\nu}^a)^2 + \frac{1}{4g_1^2}(F_{\mu\nu})^2 \right.$$
$$\left. + |\nabla_\mu\varphi^A|^2 + \frac{g_2^2}{2}(\bar{\varphi}_A T^a \varphi^A)^2 + \frac{g_1^2}{8}(|\varphi^A|^2 - N\xi)^2 \right\}, \qquad (4.2.2)$$

while the VEV's of the squark fields (4.1.14) are

$$\langle\varphi\rangle = \sqrt{\xi}\,\mathrm{diag}\{1, 1, \ldots, 1\}. \qquad (4.2.3)$$

Since the spontaneously broken gauge U(1) is a part of the model under consideration, the model supports conventional ANO strings [36], in which one can discard the $SU(N)_{\mathrm{gauge}}$ part of the action altogether. The topological stability of the ANO string is due to the fact that $\pi_1(U(1)) = Z$.

These are not the strings we are interested in. At first sight, the triviality of the homotopy group, $\pi_1(SU(N)) = 0$, implies that there are no other topologically stable strings. This impression is false. One can combine the Z_N center of $SU(N)$ with the elements $\exp(2\pi ik/N) \in U(1)$ to get a topologically stable string solution possessing both windings, in $SU(N)$ and U(1). In other words,

$$\pi_1\big(SU(N) \times U(1)/Z_N\big) \neq 0. \qquad (4.2.4)$$

It is easy to see that this nontrivial topology amounts to selecting just one element of φ, say, φ^{11}, or φ^{22}, etc, and make it wind, for instance,[5]

$$\varphi_{\text{string}} = \sqrt{\xi} \, \text{diag}(1, 1, \ldots, e^{i\alpha}), \quad x \to \infty. \tag{4.2.5}$$

Such strings can be called elementary; their tension is $1/N$ th of that of the ANO string. The ANO string can be viewed as a bound state of N elementary strings.

More concretely, one of the Z_N string solutions (a progenitor of the non-Abelian string) can be written as follows [131]:

$$\varphi = \begin{pmatrix} \phi_2(r) & 0 & \cdots & 0 \\ \cdots & \cdots & \cdots & \cdots \\ 0 & \cdots & \phi_2(r) & 0 \\ 0 & 0 & \cdots & e^{i\alpha}\phi_1(r) \end{pmatrix},$$

$$A_i^{\text{SU}(N)} = \frac{1}{N} \begin{pmatrix} 1 & \cdots & 0 & 0 \\ \cdots & \cdots & \cdots & \cdots \\ 0 & \cdots & 1 & 0 \\ 0 & 0 & \cdots & -(N-1) \end{pmatrix} (\partial_i\alpha)\big[-1 + f_{NA}(r)\big],$$

$$A_i^{\text{U}(1)} = \frac{I}{2} A_i = \frac{I}{N} (\partial_i\alpha)\big[1 - f(r)\big], \quad A_0^{\text{U}(1)} = A_0^{\text{SU}(N)} = 0, \tag{4.2.6}$$

where $i = 1, 2$ labels the coordinates in the plane orthogonal to the string axis, r and α are the polar coordinates in this plane, and I is the unit $N \times N$ matrix. Other Z_N string solutions are obtained by permutations of the rotating flavor.

The profile functions $\phi_1(r)$ and $\phi_2(r)$ determine the profiles of the scalar fields, while $f_{NA}(r)$ and $f(r)$ determine the SU(N) and U(1) fields of the string solutions, respectively. These functions satisfy the following rather obvious boundary conditions:

$$\phi_1(0) = 0,$$
$$f_{NA}(0) = 1, \quad f(0) = 1, \tag{4.2.7}$$

at $r = 0$, and

$$\phi_1(\infty) = \sqrt{\xi}, \quad \phi_2(\infty) = \sqrt{\xi},$$
$$f_{NA}(\infty) = 0, \quad f(\infty) = 0 \tag{4.2.8}$$

at $r = \infty$.

---

[5] As explained below, α is the angle of the coordinate $\vec{x}_\perp$ in the perpendicular plane.

Now, let us derive the first-order equations which determine the profile functions, making use of the Bogomol'nyi representation [5] of the model (4.2.2). We have

$$T = \int d^2x \left\{ \left[\frac{1}{\sqrt{2g_2}} F_3^{*a} + \frac{g_2}{\sqrt{2}} (\bar{\varphi}_A T^a \varphi^A) \right]^2 \right.$$

$$+ \left[\frac{1}{\sqrt{2g_1}} F_3^* + \frac{g_1}{2\sqrt{2}} (|\varphi^A|^2 - N\xi) \right]^2$$

$$\left. + |\nabla_1 \varphi^A + i\nabla_2 \varphi^A|^2 + \frac{N}{2} \xi F_3^* \right\}, \qquad (4.2.9)$$

where

$$F_3^* = F_{12} \quad \text{and} \quad F_3^{*a} = F_{12}^a,$$

and we assume that the fields in question depend only on the transverse coordinates $x_i, i = 1, 2$.

The Bogomol'nyi representation (4.2.9) leads us to the following first-order equations:

$$F_3^* + \frac{g_1^2}{2} (|\varphi^A|^2 - N\xi) = 0,$$

$$F_3^{*a} + g_2^2 (\bar{\varphi}_A T^a \varphi^A) = 0,$$

$$(\nabla_1 + i\nabla_2)\varphi^A = 0. \qquad (4.2.10)$$

Once these equations are satisfied, the energy of the BPS object is given by the last surface term in (4.2.9). Note that the representation (4.2.9) can be written also with the opposite sign in front of the flux terms. Then we would get the Bogomol'nyi equations for the anti-string.

For minimal winding we substitute the *ansatz* (4.2.6) in Eqs. (4.2.10) to get the first-order equations for the profile functions of the Z_N string [140, 131],

$$r \frac{d}{dr} \phi_1(r) - \frac{1}{N} \left(f(r) + (N-1) f_{NA}(r) \right) \phi_1(r) = 0,$$

$$r \frac{d}{dr} \phi_2(r) - \frac{1}{N} \left(f(r) - f_{NA}(r) \right) \phi_2(r) = 0,$$

$$-\frac{1}{r} \frac{d}{dr} f(r) + \frac{g_1^2 N}{4} \left[(N-1)\phi_2(r)^2 + \phi_1(r)^2 - N\xi \right] = 0,$$

$$-\frac{1}{r} \frac{d}{dr} f_{NA}(r) + \frac{g_2^2}{2} \left[\phi_1(r)^2 - \phi_2(r)^2 \right] = 0. \qquad (4.2.11)$$

These equations present a Z_N-string generalization of the Bogomol'nyi equations for the ANO string [5] (see also (3.2.19) and (C.13)). They were solved numerically

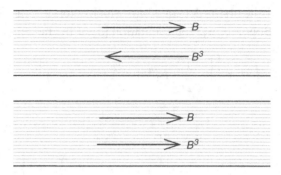

Figure 4.1. Two distinct Z_2 strings in U(2) theory.

for the U(2) case (i.e. $N = 2$) in [131]. Clearly, the solutions to the first-order equations automatically satisfy the second-order equations of motion.

The tension of this elementary Z_N string is

$$T_1 = 2\pi\,\xi. \qquad (4.2.12)$$

Since our string is a BPS object, this result is exact and has neither perturbative nor nonperturbative corrections. Note that the tension of the ANO string is N times larger; in our normalization

$$T_{\mathrm{ANO}} = 2\pi\,N\,\xi. \qquad (4.2.13)$$

Clearly, the *ansatz* (4.2.6) admits permutations, leading to other Z_N string solutions of type (4.2.6). They can be obtained by changing the position of the "winding" field in Eq. (4.2.6). Altogether we have N elementary Z_N strings. For instance, if $N = 2$ (i.e. the gauge group is SU(2)×U(1)), we have two distinct Z_2 strings differing by the orientation of the flux of the U(1) magnetic field with respect to that of the third isocomponent of the SU(2) magnetic field, see Fig. 4.1.

Of course, the first-order equations (4.2.11) can be also obtained using supersymmetry. We start from the supersymmetry transformations for the fermion fields in the theory (4.1.7),

$$
\begin{aligned}
\delta\lambda^{f\alpha} &= \frac{1}{2}(\sigma_\mu\bar{\sigma}_\nu\epsilon^{\,f})^\alpha\,F_{\mu\nu} + \epsilon^{\alpha p}\,F^m(\tau^m)^f_p + \cdots, \\[4pt]
\delta\lambda^{af\alpha} &= \frac{1}{2}(\sigma_\mu\bar{\sigma}_\nu\epsilon^{\,f})^\alpha\,F^a_{\mu\nu} + \epsilon^{\alpha p}\,F^{am}(\tau^m)^f_p + \cdots, \\[4pt]
\delta\bar{\tilde{\psi}}^{kA}_{\dot\alpha} &= i\sqrt{2}\,\bar{\nabla}_{\dot\alpha\alpha}q^{kA}_f\epsilon^{\alpha f} + \cdots, \\[4pt]
\delta\bar{\psi}_{\dot\alpha Ak} &= i\sqrt{2}\,\bar{\nabla}_{\dot\alpha\alpha}\bar{q}_{fAk}\epsilon^{\alpha f} + \cdots
\end{aligned}
\qquad (4.2.14)
$$

Here $f = 1, 2$ is the SU(2)$_R$ index and $\lambda^{f\alpha}$ and $\lambda^{af\alpha}$ are the fermions from the $\mathcal{N} = 2$ vector supermultiplets of the U(1) and SU(2) factors, respectively, while q^{kAf} denotes the SU(2)$_R$ doublet of the squark fields q^{kA} and $\bar{\tilde{q}}^{Ak}$ in the quark hypermultiplets. The parameters of the SUSY transformations in the microscopic theory are denoted as $\epsilon^{\alpha f}$. Furthermore, the F terms in Eq. (4.2.14) are

$$F^1 + iF^2 = i\,\frac{g_1^2}{2}\,(\mathrm{Tr}\,|\varphi|^2 - N\xi), \quad F^3 = 0 \qquad (4.2.15)$$

for the U(1) field, and

$$F^{a1} + iF^{a2} = i\,g_2^2\,\mathrm{Tr}\,(\bar{\varphi}T^a\varphi), \quad F^{a3} = 0 \qquad (4.2.16)$$

for the SU(N) field. The dots in (4.2.14) stand for terms involving the adjoint scalar fields which vanish on the string solution (in the equal mass case) because the adjoint fields are given by their vacuum expectation values (4.1.11).

In Ref. [35] it was shown that four supercharges selected by the conditions

$$\epsilon^{12} = -\epsilon^{11}, \quad \epsilon^{21} = \epsilon^{22} \qquad (4.2.17)$$

act trivially on the BPS string. Imposing the conditions (4.2.17) and requiring the left-hand sides of Eqs. (4.2.14) to vanish[6] we get, upon substituting the *ansatz* (4.2.6), the first-order equations (4.2.11).

---

[6] If, instead of (4.2.17), we required other combinations of the SUSY transformation parameters to vanish (changing the signs in (4.2.17)) we would get the anti-string equations, with the opposite direction of the gauge fluxes.

4.3 Elementary non-Abelian strings

The elementary Z_N strings in the model (4.1.7) give rise to non-Abelian strings provided the condition (4.1.13) is satisfied [130, 131, 132, 133]. This means that, in addition to trivial translational moduli, they have extra moduli corresponding to spontaneous breaking (on the string) of a non-Abelian symmetry acting in the bulk theory in the infrared. Indeed, while the "flat" vacuum (4.1.14) is $SU(N)_{C+F}$ symmetric, the solution (4.2.6) breaks this symmetry[7] down to $U(1)\times SU(N-1)$ (at $N > 2$). This ensures the presence of $2(N-1)$ orientational moduli.

To obtain the non-Abelian string solution from the Z_N string (4.2.6) we apply the diagonal color-flavor rotation preserving the vacuum (4.1.14). To this end it is convenient to pass to the singular gauge where the scalar fields have no winding at infinity, while the string flux comes from the vicinity of the origin. In this gauge we have

$$
\varphi = U
\begin{pmatrix}
\phi_2(r) & 0 & \ldots & 0 \\
\ldots & \ldots & \ldots & \ldots \\
0 & \ldots & \phi_2(r) & 0 \\
0 & 0 & \ldots & \phi_1(r)
\end{pmatrix}
U^{-1},
$$

$$
A_i^{SU(N)} = \frac{1}{N} U
\begin{pmatrix}
1 & \ldots & 0 & 0 \\
\ldots & \ldots & \ldots & \ldots \\
0 & \ldots & 1 & 0 \\
0 & 0 & \ldots & -(N-1)
\end{pmatrix}
U^{-1} \left(\partial_i \alpha\right) f_{NA}(r),
$$

$$
A_i^{U(1)} = -\frac{1}{N} \left(\partial_i \alpha\right) f(r), \quad A_0^{U(1)} = A_0^{SU(N)} = 0, \tag{4.3.1}
$$

where U is a matrix $\in SU(N)_{C+F}$. This matrix parametrizes orientational zero modes of the string associated with flux rotation in $SU(N)$. Since the diagonal color-flavor symmetry is not broken by the VEV's of the scalar fields in the bulk (color-flavor locking) it is physical and has nothing to do with the gauge rotations eaten by the Higgs mechanism. The orientational moduli encoded in the matrix U are *not* gauge artifacts.

The orientational zero modes of a non-Abelian string were first observed in [130, 131]. In Ref. [130] a general index theorem was proved which shows that the dimension of elementary string moduli space is $2N = 2(N-1) + 2$ where 2 stands for translational moduli while $2(N-1)$ is the dimension of the internal moduli space.[8] In Ref. [131] the explicit solution for the non-Abelian string which we review here was found and explored.

---

[7] At $N = 2$ the string solution breaks SU(2) down to U(1).
[8] The index theorem in [130] deals with more general multiple strings. It was shown that the dimension of the moduli space of the k-string solution is $2kN$.

In fact, non-translational zero modes of strings were discussed earlier in a U(1)×U(1) model [149, 150], and somewhat later, in more contrived models, in Ref. [151]. (The latter paper is entitled "Zero modes of non-Abelian vortices"!) It is worth emphasizing that, along with some apparent similarities, there are drastic distinctions between the "non-Abelian strings" we review here and the strings that were discussed in the 1980s. In particular, in the example treated in Ref. [151] the gauge group is not completely broken in the vacuum, and, therefore, there are massless gauge fields in the bulk. If the unbroken generator acts non trivially on the string flux (which is proportional to a broken generator) then it can and does create zero modes. Infrared divergence problems ensue immediately.

In the case we treat here the gauge group is completely broken (up to a discrete subgroup Z_N). The theory in the bulk is fully Higgsed. The unbroken group $SU(N)_{C+F}$, a combination of the gauge and flavor groups, is global. There are no massless fields in the bulk.

It is possible to model the example considered in [151] if we gauge the unbroken global symmetry $SU(N)_{C+F}$ of the model (4.1.7) with respect to yet another gauge field B_μ.

Let us also note that a generalization of the non-Abelian string solutions in six-dimensional gauge theory with eight supercharges was carried out in [152] while the non-Abelian strings in strongly coupled vacua were considered in [153].

4.4 The world-sheet effective theory

The non-Abelian string solution (4.3.1) is characterized by two translational moduli (the position of the string in the (1,2) plane) and $2(N-1)$ orientational moduli. Below we review the effective two-dimensional low-energy theory on the string world sheet. As usual, the translational moduli decouple and we focus on the internal dynamics of the orientational moduli. Our string is a 1/2-BPS state in $\mathcal{N} = 2$

supersymmetric gauge theory with eight supercharges. Thus it has four supercharges acting in the world sheet theory. This means that we have extended $\mathcal{N} = 2$ supersymmetric effective theory on the string world sheet. This theory turns out to be a two-dimensional CP($N - 1$) model [130, 131, 132, 133]. In Section 4.4 we will first present a derivation of this theory and then discuss the underlying physics.

4.4.1 Derivation of the CP($N - 1$) model

Now, following Refs. [131, 132, 154], we will derive the effective low-energy theory for the moduli residing in the matrix U in the problem at hand. As is clear from the string solution (4.3.1), not each element of the matrix U will give rise to a modulus. The SU($N - 1$)×U(1) subgroup remains unbroken by the string solution under consideration; therefore the moduli space is

$$\frac{\text{SU}(N)}{\text{SU}(N - 1) \times \text{U}(1)} \sim \text{CP}(N - 1). \tag{4.4.1}$$

Keeping this in mind we parametrize the matrices entering Eq. (4.3.1) as follows:

$$\frac{1}{N}\left\{U\begin{pmatrix} 1 & \cdots & 0 & 0 \\ \cdots & \cdots & \cdots & \cdots \\ 0 & \cdots & 1 & 0 \\ 0 & 0 & \cdots & -(N - 1) \end{pmatrix}U^{-1}\right\}_p^l = -n^l n_p^* + \frac{1}{N}\delta_p^l, \tag{4.4.2}$$

where n^l is a complex vector in the fundamental representation of SU(N), and

$$n_l^* n^l = 1, \tag{4.4.3}$$

($l, p = 1, \ldots, N$ are color indices). As we will show below, one U(1) phase will be gauged away in the effective sigma model. This gives the correct number of degrees of freedom, namely, $2(N - 1)$.

With this parametrization the string solution (4.3.1) can be rewritten as

$$\varphi = \frac{1}{N}[(N - 1)\phi_2 + \phi_1] + (\phi_1 - \phi_2)\left(n \cdot n^* - \frac{1}{N}\right),$$

$$A_i^{\text{SU}(N)} = \left(n \cdot n^* - \frac{1}{N}\right)\varepsilon_{ij}\frac{x_j}{r^2}f_{NA}(r),$$

$$A_i^{\text{U}(1)} = \frac{1}{N}\varepsilon_{ij}\frac{x_j}{r^2}f(r), \tag{4.4.4}$$

where for brevity we suppress all SU(N) indices. The notation is self-evident.

Assume that the orientational moduli are slowly varying functions of the string world-sheet coordinates $x_k, k = 0, 3$. Then the moduli n^l become fields of a $(1+1)$-dimensional sigma model on the world sheet. Since n^l parametrize the string zero modes, there is no potential term in this sigma model.

To obtain the kinetic term we substitute our solution (4.4.4), which depends on the moduli n^l, in the action (4.2.2), assuming that the fields acquire a dependence on the coordinates x_k via $n^l(x_k)$. In doing so we immediately observe that we have to modify our solution: we have to include in it the $k = 0, 3$ components of the gauge potential which are no longer vanishing. In the CP(1) case, as was shown in [132], the potential A_k must be orthogonal (in the SU(2) space) to the matrix (4.4.2), as well as to its derivatives with respect to x_k. Generalization of these conditions to the CP$(N-1)$ case leads to the following *ansatz*:

$$A_k^{SU(N)} = -i\left[\partial_k n \cdot n^* - n \cdot \partial_k n^* - 2n \cdot n^*(n^*\partial_k n)\right]\rho(r), \quad \alpha = 0, 3, \quad (4.4.5)$$

where we assume the contraction of the color indices inside the parentheses,

$$(n^*\partial_k n) \equiv n_l^* \partial_k n^l,$$

and introduce a new profile function $\rho(r)$.

The function $\rho(r)$ in Eq. (4.4.5) is determined through a minimization procedure [131, 132, 154] which generates ρ's own equation of motion. Now we will outline its derivation. But at first we note that $\rho(r)$ vanishes at infinity,

$$\rho(\infty) = 0. \quad (4.4.6)$$

The boundary condition at $r = 0$ will be determined shortly.

The kinetic term for n^l comes from the gauge and quark kinetic terms in Eq. (4.2.2). Using Eqs. (4.4.4) and (4.4.5) to calculate the SU(N) gauge field strength we find

$$F_{ki}^{SU(N)} = \left(\partial_k n \cdot n^* + n \cdot \partial_k n^*\right)\varepsilon_{ij}\frac{x_j}{r^2}f_{NA}\left[1 - \rho(r)\right]$$

$$+ i\left[\partial_k n \cdot n^* - n \cdot \partial_k n^* - 2n \cdot n^*(n^*\partial_k n)\right]\frac{x_i}{r}\frac{d\rho(r)}{dr}. \quad (4.4.7)$$

In order to have a finite contribution from the term Tr F_{ki}^2 in the action we have to impose the constraint

$$\rho(0) = 1. \quad (4.4.8)$$

Substituting the field strength (4.4.7) in the action (4.2.2) and including, in addition, the quark kinetic term, after rather straightforward but tedious algebra we arrive at

$$S^{(1+1)} = 2\beta \int dt\, dz \left\{ (\partial_k n^* \partial_k n) + (n^* \partial_k n)^2 \right\},$$ (4.4.9)

where the coupling constant β is given by

$$\beta = \frac{2\pi}{g_2^2} I,$$ (4.4.10)

and I is a basic normalizing integral

$$I = \int_0^\infty r\, dr \left\{ \left(\frac{d}{dr} \rho(r) \right)^2 + \frac{1}{r^2} f_{NA}^2 \left(1 - \rho \right)^2 \right.$$
$$\left. + g_2^2 \left[\frac{\rho^2}{2} (\phi_1^2 + \phi_2^2) + (1 - \rho)(\phi_2 - \phi_1)^2 \right] \right\}.$$ (4.4.11)

The theory in Eq. (4.4.9) is nothing but the two-dimensional CP($N - 1$) model. To see that this is indeed the case we can eliminate the second term in (4.4.9) introducing a non-propagating U(1) gauge field. We review this in Section 4.4.3 (see also Appendix B), and then discuss the underlying physics of the model.

Thus, we obtain the CP($N - 1$) model as an effective low-energy theory on the world sheet of the non-Abelian string. Its coupling constant β is related to the four-dimensional coupling g_2^2 via the basic normalizing integral (4.4.11). This integral must be viewed as an "action" for the profile function ρ.

Varying (4.4.11) with respect to ρ one obtains the second-order equation which the function ρ must satisfy, namely,

$$-\frac{d^2}{dr^2} \rho - \frac{1}{r} \frac{d}{dr} \rho - \frac{1}{r^2} f_{NA}^2 (1 - \rho) + \frac{g_2^2}{2} (\phi_1^2 + \phi_2^2)\rho - \frac{g_2^2}{2} (\phi_1 - \phi_2)^2 = 0.$$ (4.4.12)

After some algebra and extensive use of the first-order equations (4.2.11) one can show that the solution of (4.4.12) is

$$\rho = 1 - \frac{\phi_1}{\phi_2}.$$ (4.4.13)

This solution satisfies the boundary conditions (4.4.6) and (4.4.8). Substituting this solution back in the expression for the normalizing integral (4.4.11) one can check that this integral reduces to a total derivative and is given by the flux of the string determined by $f_{NA}(0) = 1$. In this way we arrive at

$$I = 1. \tag{4.4.14}$$

This result can be traced back to the fact that our theory (4.2.2) is $\mathcal{N} = 2$ supersymmetric theory, and the string is BPS saturated. In Section 4.5 we will see that this fact is crucial for the interpretation of confined monopoles as sigma-model kinks. Generally speaking, for non-BPS strings, I could be a certain function of N (see Ref. [155] for a particular example).

Equation (4.4.14) implies

$$\beta = \frac{2\pi}{g_2^2}. \tag{4.4.15}$$

The two-dimensional coupling is determined by the four-dimensional non-Abelian coupling. This relation is obtained at the classical level. In quantum theory both couplings run. Therefore, we have to specify a scale at which the relation (4.4.15) takes place. The two-dimensional CP($N - 1$) model (4.4.9) is an effective low-energy theory appropriate for the description of internal string dynamics at low energies, lower than the inverse thickness of the string which is given by the masses of the gauge/quark multiplets (4.1.16) and (4.1.17) in the bulk SU(N)×U(1) theory. Thus, the parameter $g\sqrt{\xi}$ plays the role of a physical ultraviolet (UV) cut off in the action (4.4.9). This is the scale at which Eq. (4.4.15) holds. Below this scale, the coupling β runs according to its two-dimensional renormalization-group flow, see Section 4.4.3. It is worth noting that if the bulk theory were not Higgsed the running law of the bulk theory would exactly match that of the CP($N - 1$) model. Indeed, Eq. (4.4.15) implies that

$$\left(4\pi\beta_{\mathrm{CP}(N-1)}\right)_0 = \left(\frac{8\pi^2}{g_2^2}\right)_0. \tag{4.4.16}$$

To get the running couplings we must add $-b_{\mathrm{CP}(N-1)} \ln(M_0/\mu)$ on the left-hand side of Eq. (4.4.16) and $-b_{\mathrm{SU}(N)\times\mathrm{U}(1)} \ln(M_0/\mu)$ on the right-hand side. The coefficients of the two- and four-dimensional Gell-Mann–Low functions coincide, $b_{\mathrm{CP}(N-1)} = b_{\mathrm{SU}(N)\times\mathrm{U}(1)} = N$.

Thus – we repeat again – the model (4.4.9) describes the low-energy limit: all higher-order terms in derivatives are neglected. Quartic in derivatives, sextic, and so on, terms certainly exist. In fact, the derivative expansion runs in powers of

$$\left(g_2 \sqrt{\xi}\right)^{-1} \partial_\alpha, \tag{4.4.17}$$

where $g_2\sqrt{\xi}$ gives the order of magnitude of masses in the bulk theory. The sigma model (4.4.9) is adequate at scales below $g_2\sqrt{\xi}$ where the higher-derivative corrections are negligibly small.

To conclude this section let us narrow down the model (4.4.9) setting $N = 2$. In this case we deal with the CP(1) model equivalent to the O(3) sigma model. The action (4.4.9) can be represented as (see Appendix B)

$$S^{(1+1)} = \frac{\beta}{2} \int dt\, dz\, (\partial_k S^a)^2, \tag{4.4.18}$$

where S^a $(a = 1, 2, 3)$ is a real unit vector, $(S^a)^2 = 1$, sweeping the two-dimensional sphere S_2. It is defined as

$$S^a = -n^* \tau^a n. \tag{4.4.19}$$

The model (4.4.18), as an effective theory on the world sheet of the non-Abelian string in SU(2)×U(1) SQCD with $\mathcal{N} = 2$ supersymmetry, was first derived in [131] in a field-theoretical framework. This derivation was generalized for arbitrary N in [154]. A brane construction of (4.4.9) was presented in [130].

4.4.2 Fermion zero modes

In Section 4.4.1 we derived the bosonic part of the effective $\mathcal{N} = 2$ supersymmetric CP($N-1$) model. Now we will find fermion zero modes for the non-Abelian string. Inclusion of these modes into consideration will demonstrate that the internal world sheet dynamics is given by $\mathcal{N} = 2$ supersymmetric CP($N-1$) model. This program was carried out in [132] for $N = 2$. Here we will focus on this construction.

The string solution (4.4.4) in the SU(2)×U(1) theory reduces to

$$\varphi = U \begin{pmatrix} \phi_2(r) & 0 \\ 0 & \phi_1(r) \end{pmatrix} U^{-1},$$

$$A_i^a(x) = -S^a\, \varepsilon_{ij}\, \frac{x_j}{r^2}\, f_{NA}(r),$$

$$A_i(x) = \varepsilon_{ij}\, \frac{x_j}{r^2}\, f(r), \tag{4.4.20}$$

while the parametrization (4.4.2) reduces to

$$S^a \tau^a = U \tau^3 U^{-1}, \quad a = 1, 2, 3, \tag{4.4.21}$$

by virtue of Eq. (4.4.19).

Our string solution is 1/2 BPS-saturated. This means that four supercharges, out of eight of the four-dimensional theory (4.1), act trivially on the string solution

(4.4.20). The remaining four supercharges generate four fermion zero modes which were termed supertranslational modes because they are superpartners to two translational zero modes. The corresponding four fermionic moduli are superpartners to the coordinates x_0 and y_0 of the string center. The supertranslational fermion zero modes were found in Ref. [35] for the U(1) ANO string in $\mathcal{N} = 2$ theory. This is discussed in detail in Appendix C, see Section C.3. Transition to the non-Abelian model at hand is absolutely straightforward. We will not dwell on this procedure here.

Instead, we will focus on four *additional* fermion zero modes which arise only for the non-Abelian string, to be referred to as superorientational. They are superpartners of the bosonic orientational moduli S^a.

Let us see how one can explicitly construct these four zero modes (in CP(1)) and study their impact on the string world sheet.

At $\mathcal{N} = 2$ the fermionic part of the action of the model (4.1.7) is

$$
\begin{aligned}
S_{\text{ferm}} = \int d^4x \Bigg\{ & \frac{i}{g_2^2} \bar{\lambda}_f^a \slashed{D} \lambda^{af} + \frac{i}{g_1^2} \bar{\lambda}_f \slashed{\partial} \lambda^f + \text{Tr}\left[\bar{\psi} i \slashed{\nabla} \psi \right] + \text{Tr}\left[\tilde{\psi} i \slashed{\nabla} \bar{\tilde{\psi}} \right] \\
& + \frac{1}{\sqrt{2}} \varepsilon^{abc} \bar{a}^a (\lambda_f^b \lambda^{cf}) + \frac{1}{\sqrt{2}} \varepsilon^{abc} (\bar{\lambda}^{bf} \bar{\lambda}_f^c) a^c \\
& + \frac{i}{\sqrt{2}} \text{Tr}\left[\bar{q}_f (\lambda^f \psi) + (\tilde{\psi} \lambda_f) q^f + (\bar{\psi} \bar{\lambda}_f) q^f + \bar{q}^f (\bar{\lambda}_f \bar{\tilde{\psi}}) \right] \\
& + \frac{i}{\sqrt{2}} \text{Tr}\left[\bar{q}_f \tau^a (\lambda^{af} \psi) + (\tilde{\psi} \lambda_f^a) \tau^a q^f + (\bar{\psi} \bar{\lambda}_f^a) \tau^a q^f + \bar{q}^f \tau^a (\bar{\lambda}_f^a \bar{\tilde{\psi}}) \right] \\
& + \frac{i}{\sqrt{2}} \text{Tr}\left[\bar{\psi} \left(a + a^a \tau^a \right) \psi \right] + \frac{i}{\sqrt{2}} \text{Tr}\left[\tilde{\psi} \left(a + a^a \tau^a \right) \bar{\tilde{\psi}} \right] \Bigg\}, \quad (4.4.22)
\end{aligned}
$$

where we use the matrix color-flavor notation for the matter fermions $(\psi^\alpha)^{kA}$ and $(\tilde{\psi}^\alpha)_{Ak}$. The traces in Eq. (4.4.22) are performed over the color-flavor indices. Contraction of spinor indices is assumed inside the parentheses, say, $(\lambda\psi) \equiv \lambda_\alpha \psi^\alpha$.

As was mentioned in Section 4.2, the four supercharges selected by the conditions (4.2.17) act trivially on the BPS string in the theory with the FI term of the F type. To generate superorientational fermion zero modes the following method was used in [132]. Assume the orientational moduli S^a in the string solution (4.4.20) to have a slow dependence on the world-sheet coordinates x_0 and x_3 (or t and z). Then the four (real) supercharges selected by the conditions (4.2.17) no longer act trivially. Instead, their action now generates fermion fields proportional to x_0 and x_3 derivatives of S^a.

This is exactly what one expects from the residual $\mathcal{N} = 2$ supersymmetry in the world sheet theory. The above four supercharges generate the world-sheet supersymmetry in the $\mathcal{N} = 2$ two-dimensional CP(1) model,

$$\delta \chi_1^a = i\sqrt{2}\Big[(\partial_0 + i\partial_3)\, S^a\, \varepsilon_2 + \varepsilon^{abc} S^b\, (\partial_0 + i\partial_3)\, S^c\, \eta_2\Big],$$

$$\delta \chi_2^a = i\sqrt{2}\Big[(\partial_0 - i\partial_3)\, S^a\, \varepsilon_1 + \varepsilon^{abc} S^b\, (\partial_0 - i\partial_3)\, S^c\, \eta_1\Big]. \qquad (4.4.23)$$

Here χ_α^a ($\alpha = 1, 2$ is the spinor index) are real two-dimensional fermions of the CP(1) model. They are superpartners of S^a and are subject to the orthogonality condition (see Appendix B)

$$S^a \chi_\alpha^a = 0. \qquad (4.4.24)$$

The real parameters of the $\mathcal{N} = 2$ two-dimensional SUSY transformations ε_α and η_α are identified with the parameters of the four-dimensional SUSY transformations (with the constraint (4.2.17)) as follows:

$$\varepsilon_1 - i\eta_1 = \frac{1}{\sqrt{2}}(\epsilon^{21} + \epsilon^{22}) = \sqrt{2}\epsilon^{22},$$

$$\varepsilon_2 + i\eta_2 = \frac{1}{\sqrt{2}}(\epsilon^{11} - \epsilon^{12}) = \sqrt{2}\epsilon^{11}. \qquad (4.4.25)$$

In this way the world-sheet supersymmetry was used to re-express the fermion fields obtained upon the action of these four supercharges in terms of the $(1 + 1)$-dimensional fermions. This procedure gives us the superorientational fermion zero modes [132],

$$\bar{\psi}_{Ak\dot{2}} = \left(\frac{\tau^a}{2}\right)_{Ak} \frac{1}{2\phi_2}(\phi_1^2 - \phi_2^2)\Big[\chi_2^a + i\varepsilon^{abc} S^b \chi_2^c\Big],$$

$$\bar{\psi}_{\dot{1}}^{kA} = \left(\frac{\tau^a}{2}\right)^{kA} \frac{1}{2\phi_2}(\phi_1^2 - \phi_2^2)\Big[\chi_1^a - i\varepsilon^{abc} S^b \chi_1^c\Big],$$

$$\bar{\psi}_{Ak\dot{1}} = 0, \quad \bar{\psi}_{\dot{2}}^{kA} = 0,$$

$$\lambda^{a22} = \frac{i}{2}\frac{x_1 + ix_2}{r^2} f_{NA}\frac{\phi_1}{\phi_2}\Big[\chi_1^a - i\varepsilon^{abc} S^b \chi_1^c\Big],$$

$$\lambda^{a11} = \frac{i}{2}\frac{x_1 - ix_2}{r^2} f_{NA}\frac{\phi_1}{\phi_2}\Big[\chi_2^a + i\varepsilon^{abc} S^b \chi_2^c\Big],$$

$$\lambda^{a12} = \lambda^{a11}, \quad \lambda^{a21} = \lambda^{a22}, \qquad (4.4.26)$$

where the dependence on x_i is encoded in the string profile functions, see Eq. (4.4.20).

Now let us directly check that the zero modes (4.4.26) satisfy the Dirac equations of motion. From the fermion action of the model (4.4.22) we get the Dirac equations for λ^a,

$$\frac{i}{g_2^2} \bar{\slashed{D}} \lambda^{af} + \frac{i}{\sqrt{2}} \operatorname{Tr}\left(\bar{\psi} \tau^a q^f + \bar{q}^f \tau^a \tilde{\psi} \right) = 0. \tag{4.4.27}$$

At the same time, for the matter fermions,

$$i \slashed{\nabla} \bar{\psi} + \frac{i}{\sqrt{2}} \left[\bar{q}_f \lambda^f - (\tau^a \bar{q}_f) \lambda^{af} + (a - a^a \tau^a) \tilde{\psi} \right] = 0,$$

$$i \slashed{\nabla} \bar{\tilde{\psi}} + \frac{i}{\sqrt{2}} \left[\lambda_f q^f + \lambda_f^a (\tau^a q^f) + (a + a^a \tau^a) \psi \right] = 0. \tag{4.4.28}$$

Next, we substitute the orientational fermion zero modes (4.4.26) into these equations. After some algebra one can check that (4.4.26) do satisfy the Dirac equations (4.4.27) and (4.4.28) provided the first-order equations for the string profile functions (4.2.11) are satisfied.

Furthermore, it is instructive to check that the zero modes (4.4.26) do produce the fermion part of the $\mathcal{N} = 2$ two-dimensional CP(1) model. To this end we return to the usual assumption that the fermion collective coordinates χ_α^a in Eq. (4.4.26) have an adiabatic dependence on the world-sheet coordinates x_k ($k = 0, 3$). This is quite similar to the procedure of Section 4.4.1 for bosonic moduli. Substituting Eq. (4.4.26) in the fermion kinetic terms in the bulk theory (4.4.22), and taking into account the derivatives of χ_α^a with respect to the world-sheet coordinates, we arrive at

$$\beta \int dt\, dz \left\{ \frac{1}{2} \chi_1^a (\partial_0 - i\partial_3) \chi_1^a + \frac{1}{2} \chi_2^a (\partial_0 + i\partial_3) \chi_2^a \right\}, \tag{4.4.29}$$

where β is given by the same integral (4.4.15) as for the bosonic kinetic term, see Eq. (4.4.18).

Finally we must discuss the four-fermion interaction term in the CP(1) model. We can use the world sheet $\mathcal{N} = 2$ supersymmetry to reconstruct this term. The SUSY transformations in the CP(1) model have the form (see e.g. [156] for a review)

$$\delta \chi_1^a = i\sqrt{2}\, (\partial_1 + i\partial_3)\, S^a\, \varepsilon_2 - \sqrt{2}\varepsilon_1\, S^a (\chi_1^b \chi_2^b),$$

$$\delta \chi_2^a = i\sqrt{2}\, (\partial_1 - i\partial_3)\, S^a\, \varepsilon_1 + \sqrt{2}\varepsilon_2\, S^a (\chi_1^b \chi_2^b),$$

$$\delta S^a = \sqrt{2}(\varepsilon_1 \chi_2^a + \varepsilon_2 \chi_1^a), \tag{4.4.30}$$

where we put $\eta_\alpha = 0$ for simplicity. Imposing this supersymmetry leads to the following effective theory on the string world sheet

$$S_{CP(1)} = \beta \int dt dz \left\{ \frac{1}{2} (\partial_k S^a)^2 + \frac{1}{2} \chi_1^a i (\partial_0 - i \partial_3) \chi_1^a \right.$$

$$\left. + \frac{1}{2} \chi_2^a i (\partial_0 + i \partial_3) \chi_2^a - \frac{1}{2} (\chi_1^a \chi_2^a)^2 \right\}, \tag{4.4.31}$$

This is indeed the action of the $\mathcal{N} = 2$ CP(1) sigma model in its entirety.

4.4.3 Physics of the CP($N - 1$) model with $\mathcal{N} = 2$

As is quite common in two dimensions, the Lagrangian of our effective theory on the string world sheet can be cast in many different (but equivalent) forms. In particular, the $\mathcal{N} = 2$ supersymmetric CP($N - 1$) model (4.4.9) can be understood as a strong-coupling limit of a U(1) gauge theory [157]. Then the bosonic part of the action takes the form

$$S = \int d^2 x \left\{ 2\beta |\nabla_k n^\ell|^2 + \frac{1}{4e^2} F_{kl}^2 + \frac{1}{e^2} |\partial_k \sigma|^2 \right.$$

$$\left. + 4\beta |\sigma|^2 |n^\ell|^2 + 2e^2 \beta^2 (|n^\ell|^2 - 1)^2 \right\}, \tag{4.4.32}$$

where $\nabla_k = \partial_k - i A_k$ while σ is a complex scalar field. The condition (4.4.3) is implemented in the limit $e^2 \to \infty$. Moreover, in this limit the gauge field A_k and its $\mathcal{N} = 2$ bosonic superpartner σ become auxiliary and can be eliminated by virtue of the equations of motion,

$$A_k = -\frac{i}{2} n_\ell^* \overleftrightarrow{\partial_k} n^\ell, \quad \sigma = 0. \tag{4.4.33}$$

Substituting Eq. (4.4.33) in the Lagrangian, we can readily rewrite the action in the form (4.4.9).

The coupling constant β is asymptotically free [158]. The running coupling, as a function of energy E, is given by the formula

$$4\pi\beta = N \ln \frac{E}{\Lambda_\sigma}, \tag{4.4.34}$$

where Λ_σ is a dynamical scale of the sigma model. The ultraviolet cut off of the sigma model on the string world sheet is determined by $g_2 \sqrt{\xi}$. Equation (4.4.15) relating the two- and four-dimensional couplings is valid at this scale. Hence,

$$\Lambda_\sigma^N = g_2^N \xi^{\frac{N}{2}} e^{-\frac{8\pi^2}{g_2^2}} = \Lambda_{SU(N)}^N. \tag{4.4.35}$$

Here we take into account Eq. (4.1.20) for the dynamical scale $\Lambda_{SU(N)}$ of the SU(N) factor of the bulk theory. Note that in the bulk theory *per se*, because of the VEV's of the squark fields, the coupling constant is frozen at $g_2\sqrt{\xi}$; there are no logarithms below this scale. The logarithms of the string world-sheet theory take over. Moreover, the dynamical scales of the bulk and world-sheet theories turn out to be the same! We will explain the reason why the dynamical scale of the (1 + 1)-dimensional effective theory on the string world sheet is identical to that of the SU(N) factor of the (3 + 1)-dimensional gauge theory later, in Section 4.6.

The CP(N − 1) model was solved by Witten in the large-N limit [159]. We will briefly summarize Witten's results and translate them in terms of strings in four dimensions [132].

Classically the field n^ℓ can have arbitrary direction; therefore, one might naively expect a spontaneous breaking of SU(N) and the occurrence of massless Goldstone modes. Well, the Coleman theorem [160] teaches us that this cannot happen in two dimensions. Quantum effects restore the symmetry. Moreover, the condition (4.4.3) gets in effect relaxed. Due to strong coupling we have more degrees of freedom than in the original Lagrangian, namely all N fields n become dynamical and acquire masses Λ_σ.

As was shown by Witten [159], the model has N vacua. These N vacua differ from each other by the expectation value of the chiral bifermion operator, see e.g. [156]. At strong coupling the chiral condensate is the order parameter. The U(1) chiral symmetry of the CP(N − 1) model is explicitly broken to a discrete Z_{2N} symmetry by the chiral anomaly. The fermion condensate breaks Z_{2N} down to Z_2. That's the origin of the N-fold degeneracy of the vacuum state.

The physics of the model becomes even more transparent in the mirror representation which was established [120] for arbitrary N. In this representation one describes the CP(N − 1) model in terms of the Coulomb gas of instantons (see [161] where this was done for non-supersymmetric CP(1) model) to prove its equivalence to an affine Toda theory. The CP(N − 1) model (4.4.32) is dual to the following $\mathcal{N} = 2$ affine Toda model [120, 162, 61, 163],

$$S_{\text{mirror}} = \int d^2x\, d^2\theta\, d^2\bar\theta\, \beta^{-1} \sum_{i=1}^{N-1} \bar{Y}_i\, Y_i$$

$$+ \left\{ \Lambda_\sigma \int d^2x\, d^2\theta \left(\sum_{i=1}^{N-1} \exp\left(Y_i\right) + \prod_{i=1}^{N-1} \exp\left(-Y_i\right) \right) + \text{H.c.} \right\}. \quad (4.4.36)$$

Here the last term is a dual instanton-induced superpotential. In fact, the exact form of the kinetic term in the mirror representation is not known because it is not protected from quantum correction in β. However the superpotential in (4.4.36) is *exact*. Since the vacuum structure is entirely determined by the superpotential (4.4.36), one immediately confirms Witten's statement of N vacua.

Indeed, the scalar potential of this affine Toda theory has N minima. For example, for $N = 2$ this theory becomes $\mathcal{N} = 2$ supersymmetric sine-Gordon theory with scalar potential

$$V_{\text{SG}} = \frac{\beta}{4\pi^2} \Lambda^2_{\text{CP}(1)} \, |\sinh y|^2, \qquad (4.4.37)$$

which obviously has two minima, at $y = 0$ and $y = \pm i\pi$ (warning: the points $y = i\pi$ and $y = -i\pi$ must be identified; they present one and the same vacuum).

This mirror model explicitly exhibits a mass gap of the order of Λ_σ. It shows that there are no Goldstone bosons (corresponding to the absence of the spontaneous breaking of the SU$(N)_{C+F}$ symmetry). In terms of strings in the four-dimensional bulk theory, this means, in turn, that the magnetic flux orientation in the target space has no particular direction, it is smeared all over. The N vacua of the world-sheet theory (4.4.32) are heirs of the N "elementary" non-Abelian strings of the bulk theory. Note that these strings are in a highly quantum regime. They are *not* the Z_N strings of the quasiclassical U$(1)^{N-1}$ theory since n^ℓ is not aligned in the vacuum.

Hori and Vafa originally derived [120] the mirror representation for the CP$(N-1)$ model in the form of the Toda model. Since then other useful equivalent representations were obtained, and they were expanded to include the so-called twisted masses of which we will speak in Section 4.4.4 and subsequent sections. A particularly useful mirror representation of the twisted-mass-deformed CP$(N-1)$ model was exploited by Dorey [30].

4.4.4 Unequal quark masses

The fact that we have N distinct vacua in the world sheet theory – N distinct elementary strings – is not quite intuitive in the above consideration. This is understandable. At the classical level the $\mathcal{N} = 2$ two-dimensional CP$(N-1)$ sigma model is characterized by a continuous vacuum manifold. This is in one-to-one correspondence with continuously many strings parametrized by the moduli n^ℓ. The continuous degeneracy is lifted only after quantum effects are taken into account. These quantum effects become crucial at strong coupling. Gone with this lifting is the moduli nature of the fields n^ℓ. They become massive. This is difficult to grasp.

To make the task easier and to facilitate contact between the bulk and world sheet theories, it is instructive to start from a deformed bulk theory, so that the string moduli are lifted already at the classical level. Then the origin of the N-fold degeneracy of the non-Abelian strings becomes transparent. This will help us understand, in an intuitive manner, other features listed above. After this understanding is achieved, nothing prevents us from returning to our problem – strings with non-Abelian moduli at the classical level – by smoothly suppressing the moduli-breaking deformation. The N-fold degeneracy will remain intact as it follows from the Witten index [123].

Thus, let us drop the assumption (4.1.13) of equal mass terms and introduce small mass differences. With unequal quark masses, the U(N) gauge group is broken by the condensation of the adjoint scalars down to U(1)^N, see (4.1.11). Off-diagonal gauge bosons, as well as the off-diagonal fields of the quark matrix q^{kA}, (together with their fermion superpartners) acquire masses proportional to various mass differences $(m_A - m_B)$. The effective low-energy theory contains now only diagonal gauge and quark fields. The reduced action suitable for the search of the string solution takes the form

$$S = \int d^4x \left\{ \frac{1}{4g_2^2} \left(F_{\mu\nu}^h \right)^2 + \frac{1}{4g_1^2} \left(F_{\mu\nu} \right)^2 \right.$$

$$\left. + |\nabla_\mu \varphi^A|^2 + \frac{g_2^2}{2} \left(\bar{\varphi}_A T^h \varphi^A \right)^2 + \frac{g_1^2}{8} \left(|\varphi^A|^2 - N\xi \right)^2, \right. \quad (4.4.38)$$

where the index $h = 1, \ldots, (N-1)$ runs over the Cartan generators of the gauge group SU(N), while the matrix φ^{kA} is reduced to its diagonal components.

The same steps which previously lead us to Eqs. (4.2.10) now give the first-order string equations in the Abelian model (4.4.38),

$$F_3^* + \frac{g_1^2}{2} \left(\left| \varphi^A \right|^2 - N\xi \right) = 0,$$

$$F_3^{*h} + g_2^2 \left(\bar{\varphi}_A T^h \varphi^A \right) = 0,$$

$$(\nabla_1 + i\nabla_2)\varphi^A = 0. \quad (4.4.39)$$

As soon as the Z_N-string solutions (4.2.6) have a diagonal form, they automatically satisfy the above first-order equations.

However, the Abelian Z_N strings (4.2.6) are now *the only* solutions to these equations. The family of solutions is discrete. The global SU(N)$_{C+F}$ group is broken down to U(1)$^{N-1}$ by the mass differences, and the continuous CP($N-1$) moduli space of the non-Abelian string is lifted. In fact, the vector n^ℓ gets fixed in N possible positions,

$$n^\ell = \delta^{\ell\ell_0}, \quad \ell_0 = 1, \ldots, N. \quad (4.4.40)$$

These N solutions correspond to the Abelian Z_N strings, see (4.2.6) and (4.4.4). If the mass differences are much smaller than $\sqrt{\xi}$ the set of parameters n^ℓ becomes *quasi*moduli.

Now, our aim is to derive an effective two-dimensional theory on the string world sheet for unequal quark mass terms. With small mass differences we will still be able to introduce orientational quasimoduli n^ℓ. In terms of the effective two-dimensional theory on the string world sheet, unequal masses lead to a shallow potential for the quasimoduli n^ℓ. Let us derive this potential.

Below we will review the derivation carried out in [132] in the $SU(2) \times U(1)$ model. The case of general N is considered in [133]. In the $N = 2$ case two minima of the potential at $S = \{0, 0, \pm 1\}$ correspond to two distinct Z_2 strings.

We start from the expression for the non-Abelian string in the singular gauge (4.4.20) parametrized by the moduli S^a, and substitute it in the action (4.1.7). The only modification we actually have to make is to supplement our *ansatz* (4.4.20) by that for the adjoint scalar field a^a; the neutral scalar field a will stay fixed at its vacuum expectation value $a = -\sqrt{2}m$.

At large r the field a^a tends to its VEV aligned along the third axis in the color space,

$$\langle a^3 \rangle = -\frac{\Delta m}{\sqrt{2}}, \quad \Delta m = m_1 - m_2, \tag{4.4.41}$$

see Eq. (4.1.11). At the same time, at $r = 0$ it must be directed along the vector S^a. The reason for this behavior is easy to understand. The kinetic term for a^a in Eq. (4.1.7) contains the commutator term of the adjoint scalar and the gauge potential. The gauge potential is singular at the origin, as is seen from Eq. (4.4.20). This implies that a^a must be aligned along S^a at $r = 0$. Otherwise, the string tension would become divergent. The following *ansatz* for a^a ensures this behavior:

$$a^a = -\frac{\Delta m}{\sqrt{2}} \left[\delta^{a3} b + S^a S^3 (1 - b) \right]. \tag{4.4.42}$$

Here we introduced a new profile function $b(r)$ which, as usual, will be determined from a minimization procedure. Note that at $S^a = (0, 0, \pm 1)$ the field a^a is given by its VEV, as expected. The boundary conditions for the function $b(r)$ are

$$b(\infty) = 1, \quad b(0) = 0. \tag{4.4.43}$$

Substituting Eq. (4.4.42) in conjunction with (4.4.20) in the action (4.1.7) we get the potential

$$V_{CP(1)} = \gamma \int d^2x \, \frac{\Delta m^2}{2} (1 - S_3^2), \tag{4.4.44}$$

where γ is given by the integral

$$\gamma = \frac{2\pi}{g_2^2} \int_0^\infty r\,dr \left\{ \left(\frac{d}{dr}b(r)\right)^2 + \frac{1}{r^2} f_{NA}^2\, b^2 \right.$$

$$\left. + g_2^2 \left[\frac{1}{2}(1-b)^2 (\phi_1^2 + \phi_2^2) + b\,(\phi_1 - \phi_2)^2\right] \right\}. \qquad (4.4.45)$$

Here two first terms in the integrand come from the kinetic term of the adjoint scalar field a^a while the term in the square brackets comes from the potential in the action (4.1.7).

Minimization with respect to $b(r)$, with the constraint (4.4.43), yields

$$b(r) = 1 - \rho(r) = \frac{\phi_1}{\phi_2}(r), \qquad (4.4.46)$$

cf. Eqs. (4.4.11) and (4.4.13). Thus,

$$\gamma = I \times \frac{2\pi}{g_2^2} = \frac{2\pi}{g_2^2}. \qquad (4.4.47)$$

We see that the normalization integrals are the same for both the kinetic and the potential terms in the world-sheet sigma model, $\gamma = \beta$. As a result we arrive at the following effective theory on the string world sheet:

$$S_{CP(1)} = \beta \int d^2x \left\{\frac{1}{2}(\partial_k S^a)^2 + \frac{|\Delta m|^2}{2}(1 - S_3^2)\right\}. \qquad (4.4.48)$$

This is the only functional form that allows $\mathcal{N} = 2$ completion.[9] See also Section 3.5.

The fact that we obtain this form shows that our *ansatz* is fully adequate. The informative aspect of the procedure is (i) confirmation of the *ansatz* (4.4.42) and (ii) constructive calculation of the constant in front of $(1 - S_3^2)$ in terms of the bulk parameters. The mass-splitting parameter Δm of the bulk theory exactly coincides with the twisted mass of the world-sheet model.

The CP(1) model (4.4.48) has two vacua located at $S^a = (0, 0, \pm1)$, see Fig. 3.11. Clearly these two vacua correspond to two elementary Z_2 strings.

For generic N the potential in the CP($N-1$) model was obtained in [133]. It has the form

$$V_{CP(N-1)} = 2\beta \left\{\sum_\ell |\tilde{m}_\ell|^2 |n^\ell|^2 - \left|\sum_\ell \tilde{m}_\ell |n^\ell|^2\right|^2\right\}, \qquad (4.4.49)$$

[9] Note, that although the global SU(2)$_{C+F}$ is broken by Δm, the extended $\mathcal{N} = 2$ supersymmetry is not.

where

$$\tilde{m}_\ell = m_\ell - m, \quad m \equiv \frac{1}{N} \sum_\ell m_\ell, \quad \ell = 1, \ldots, N. \qquad (4.4.50)$$

From the perspective of the bulk theory the index ℓ of the CP($N - 1$) model coincides with the flavor index, $\ell \equiv A$. The above potential has N vacua (4.4.40) which correspond to N distinct Z_N strings in the bulk theory.

The CP($N - 1$) model with the potential (4.4.49) is nothing but a bosonic truncation of the $\mathcal{N} = 2$ two-dimensional sigma model which was termed the twisted-mass-deformed CP($N - 1$) model. This is a generalization of the massless CP($N - 1$) model which preserves four supercharges. Twisted chiral superfields in two dimensions were introduced in [32] while the twisted mass as an expectation value of the twisted chiral multiplet was suggested in [31]. CP($N - 1$) models with twisted mass were further studied in [30] and, in particular, the BPS spectra in these theories were determined exactly.

From the bulk theory standpoint the two-dimensional CP($N - 1$) model is an effective world sheet theory for the non-Abelian string, and the emergence of $\mathcal{N} = 2$ supersymmetry should be expected. As we know, the BPS nature of the strings under consideration does require the world sheet theory to have four supercharges.

The twisted-mass-deformed CP($N - 1$) model can be nicely rewritten as a strong coupling limit of a U(1) gauge theory [30]. With twisted masses of the n^ℓ fields taken into account, the bosonic part of the action (4.4.32) becomes

$$S = \int d^2x \left\{ 2\beta \, |\nabla_k n^\ell|^2 + \frac{1}{4e^2} F_{kl}^2 + \frac{1}{e^2} |\partial_k \sigma|^2 \right.$$
$$\left. + 4\beta \left| \sigma - \frac{\tilde{m}_\ell}{\sqrt{2}} \right|^2 |n^\ell|^2 + 2e^2 \beta^2 (|n^\ell|^2 - 1)^2 \right\}. \qquad (4.4.51)$$

In the limit $e^2 \to \infty$ the σ field can be excluded by virtue of an algebraic equation of motion which leads to the potential (4.4.49).

As was already mentioned, this sigma model gives an effective description of our non-Abelian string at low energies, i.e. at energies much lower than the inverse string thickness. Typical momenta in the theory (4.4.51) are of the order of $\tilde{m}$. Therefore, for the action (4.4.51) to be applicable we must impose the condition

$$|\tilde{m}_\ell| \ll g_2 \sqrt{\xi}. \qquad (4.4.52)$$

The description in terms of the twisted-mass-deformed CP($N - 1$) model gives us a much better understanding of dynamics of the non-Abelian strings. If masses $\tilde{m}_\ell$ are much larger than the scale of the CP($N - 1$) model Λ_σ, the coupling constant β

is frozen at a large scale (of the order of $\tilde{m}_\ell$) and the theory is at weak coupling. Semiclassical analysis is applicable. The theory (4.4.51) has N vacua located at

$$n^\ell = \delta^{\ell \ell_0}, \quad \sigma = \frac{\tilde{m}_{\ell_0}}{\sqrt{2}}, \quad \ell_0 = 1, \dots, N. \tag{4.4.53}$$

They correspond to the Abelian Z_N strings of the bulk theory, see (4.4.4). As we reduce the mass differences $\tilde{m}_\ell$ and hit the value Λ_σ, the CP(N − 1) model under consideration enters the strong coupling regime. At $\tilde{m}_\ell = 0$ the global SU(N)$_{C+F}$ symmetry of the bulk theory is restored. Now n^ℓ has no particular direction. The condition (4.4.3) is relaxed. Still we have N vacua in the world sheet theory (Witten's index!). They are seen in the mirror description, see Section 4.4.3. These vacua correspond to N elementary non-Abelian strings in the strong coupling quantum regime. Thus, we see that for the BPS strings the transition from the Abelian to non-Abelian regimes is smooth. As we will discuss in Chapter 5, this is not the case for non-BPS strings. In the latter case the two regimes are separated by a phase transition [154, 164].

4.5 Confined monopoles as kinks of the CP(N − 1) model

Our bulk theory (4.1.7) is in the Higgs phase and therefore the magnetic monopoles of this theory must be in the confinement phase. If we start from a theory with the SU(N + 1) gauge group broken to SU(N)×U(1) by condensation of the adjoint scalar a from which the theory (4.1.7) emerges, the monopoles of the SU(N + 1)/ SU(N) × U(1) sector can be attached to the endpoints of the Z_N strings under consideration. In the bulk theory (4.1.7) these monopoles are infinitely heavy at $m \to \infty$, and hence the Z_N strings are stable. However, the monopoles residing in the SU(N) gauge group are still present in the theory (4.1.7). As we switch on the FI parameter ξ, the squarks condense triggering confinement of these monopoles.

In this section we will show that these monopoles manifest themselves as string junctions of the non-Abelian strings and are seen as kinks in the world sheet theory interpolating between distinct vacua of the CP($N - 1$) model [165, 132, 133].

Our task in this section is to trace the evolution of the confined monopoles starting from the quasiclassical regime, and deep into the quantum regime. For illustrative purposes it will be even more instructive if we started from the limit of weakly confined monopoles, when in fact they present just slightly distorted 't Hooft–Polyakov monopoles (Fig. 4.3).

Let us start from the limit $|\Delta m_{AB}| \gg \sqrt{\xi}$ and assume all mass differences to be of the same order. In this limit the scalar quark expectation values can be neglected, and the vacuum structure is determined by VEV's of the adjoint field a^a, see (4.1.11). In the non-degenerate case the gauge symmetry SU(N) of our bulk model is broken down to U(1)$^{N-1}$ modulo possible discrete subgroups. This is the textbook situation for occurrence of the SU(N) 't Hooft–Polyakov monopoles. The monopole core size is of the order of $|\Delta m_{AB}|^{-1}$. The 't Hooft–Polyakov solution remains valid up to much larger distances, of the order of $\xi^{-1/2}$. At distances larger than $\sim \xi^{-1/2}$ the quark VEV's become important. As usual, the U(1) charge condensation leads to the formation of the U(1) magnetic flux tubes, with the transverse size of the order of $\xi^{-1/2}$ (see the upper picture in Fig. 4.3). The flux is quantized; the flux tube tension is tiny in the scale of the square of the monopole mass. Therefore, what we deal with in this limit is basically a very weakly confined 't Hooft–Polyakov monopole.

Let us verify that the confined monopole is a junction of two strings. Consider the junction of two Z_N strings corresponding to two "neighboring" vacua of the CP($N - 1$) model. For the ℓ_0-th vacuum n^ℓ is given by (4.4.53) while for the $\ell_0 + 1$-th vacuum it is given by the same equations with $\ell_0 \to \ell_0 + 1$. The flux of this junction is given by the difference of the fluxes of these two strings. Using (4.4.4) we get that the flux of the junction is

$$4\pi \times \mathrm{diag}\,\frac{1}{2}\{\ldots 0,\ 1,\ -1,\ 0,\ \ldots\}, \tag{4.5.1}$$

with the nonvanishing entries located at positions ℓ_0 and $\ell_0 + 1$. These are exactly the fluxes of $N - 1$ distinct 't Hooft–Polyakov monopoles occurring in the SU(N) gauge theory provided that SU(N) is spontaneously broken down to U(1)$^{N-1}$. For instance, in U(2) theory the junction of two Z_2 strings is shown in Fig. 4.2.

We see that in the quasiclassical limit of large $|\Delta m_{AB}|$ the Abelian monopoles play the role of junctions of the Abelian Z_N strings. Note that in various models the monopole fluxes and those of strings were shown to match each other [166, 167, 168, 169, 140, 170, 171] so that the monopoles can be confined by strings in the Higgs phase.

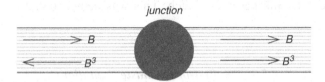

Figure 4.2. The junction of two distinct Z_2 strings in the U(2) theory.

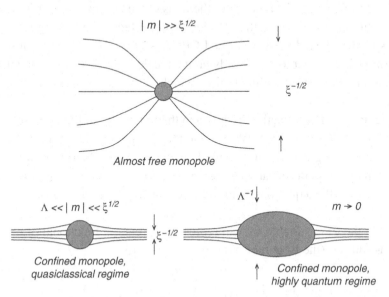

Figure 4.3. Evolution of the confined monopoles.

Now, let us reduce $|\Delta m_{AB}|$. If this parameter is limited inside the interval

$$\Lambda \ll |\Delta m_{AB}| \ll \sqrt{\xi}, \qquad (4.5.2)$$

the size of the monopole ($\sim |\Delta m_{AB}|^{-1}$) becomes larger than the transverse size of the attached strings. The monopole gets squeezed in earnest by the strings – it becomes a *bona fide* confined monopole (the lower left corner of Fig. 4.3). A natural question is how this confined monopole is seen in the effective two-dimensional CP(N−1) model (4.4.51) on the string world sheet. Since the Z_N strings of the bulk theory correspond to N vacua of the CP(N − 1) model the string junction (confined monopole) is a "domain wall" – kink – interpolating between these vacua, see Fig. 3.11.

Below we will explicitly demonstrate that in the semiclassical regime (4.5.2) the solution for the string junction in the bulk theory is in one-to-one correspondence with the kink in the world sheet theory. Then we will show that the masses of the monopole and kink perfectly match. This was demonstrated in [132] in the $N = 2$ case.

4.5.1 The first-order master equations

In this section we derive the first-order equations for the 1/4-BPS junction of the Z_N strings in the SU(N)×U(1) theory in the quasiclassical limit (4.5.2). In this limit Δm_{AB} is sufficiently small so that we can use our effective low-energy description in terms of the twisted-mass-deformed CP($N - 1$) model (4.4.51). On the other hand, Δm_{AB} is much larger than the dynamical scale of the CP($N-1$) model; hence, the latter is in the weak coupling regime which allows one to apply quasiclassical treatment.

The geometry of our junction is shown in the left corner of Fig. 4.3. Both strings are stretched along the z axis. We assume that the monopole sits near the origin, the $n^\ell = \delta^{\ell \ell_0}$-string is at negative z while the $n^\ell = \delta^{\ell(\ell_0+1)}$-string is at positive z. The perpendicular plane is parametrized by x_1 and x_2. What is sought for is a static solution of the BPS equations, with all relevant fields depending only on x_1, x_2 and z.

Ignoring the time variable we can represent the energy functional of our theory (4.2.2) as follows (the Bogomol'nyi representation [5]):

$$E = \int d^3x \left\{ \left[\frac{1}{\sqrt{2}g_2} F_3^{*a} + \frac{g_2}{2\sqrt{2}} \left(\bar{\varphi}_A \tau^a \varphi^A \right) + \frac{1}{g_2} D_3 a^a \right]^2 \right.$$

$$+ \left[\frac{1}{\sqrt{2}g_1} F_3^* + \frac{g_1}{2\sqrt{2}} \left(|\varphi^A|^2 - 2\xi \right) + \frac{1}{g_1} \partial_3 a \right]^2$$

$$+ \frac{1}{g_2^2} \left| \frac{1}{\sqrt{2}} (F_1^{*a} + i F_2^{*a}) + (D_1 + i D_2) a^a \right|^2$$

$$+ \frac{1}{g_1^2} \left| \frac{1}{\sqrt{2}} (F_1^* + i F_2^*) + (\partial_1 + i \partial_2) a \right|^2$$

$$+ \left| \nabla_1 \varphi^A + i \nabla_2 \varphi^A \right|^2$$

$$+ \left. \left| \nabla_3 \varphi^A + \frac{1}{\sqrt{2}} (a^a \tau^a + a + \sqrt{2} m_A) \varphi^A \right|^2 \right\} \qquad (4.5.3)$$

plus surface terms. As compared to the Bogomol'nyi representation (4.2.9) for strings we keep here also terms involving the adjoint fields. Following our conventions we assume the quark mass terms to be real implying that the vacuum expectation values of the adjoint scalar fields are real too. The surface terms mentioned above are

$$E_{\text{surface}} = \xi \int d^3x\, F_3^* + \sqrt{2}\,\xi \int d^2x\, \langle a \rangle \bigg|_{z=-\infty}^{z=\infty} - \sqrt{2}\,\frac{\langle a^a \rangle}{g_2^2} \int dS_n\, F_n^{*a}, \quad (4.5.4)$$

where the integral in the last term runs over a large two-dimensional sphere at $\vec{x}^2 \to \infty$. The first term on the right-hand side is related to strings, the second to domain walls, while the third to monopoles (or the string junctions).

The Bogomol'nyi representation (4.5.3) leads us to the following first-order equations:

$$F_1^* + i F_2^* + \sqrt{2}(\partial_1 + i\partial_2)a = 0,$$
$$F_1^{*a} + i F_2^{*a} + \sqrt{2}(D_1 + i D_2)a^a = 0,$$
$$F_3^* + \frac{g_1^2}{2}\left(|\varphi^A|^2 - 2\xi\right) + \sqrt{2}\,\partial_3 a = 0,$$
$$F_3^{*a} + \frac{g_2^2}{2}\left(\bar{\varphi}_A \tau^a \varphi^A\right) + \sqrt{2}\,D_3 a^a = 0,$$
$$\nabla_3 \varphi^A = -\frac{1}{\sqrt{2}}\left(a^a \tau^a + a + \sqrt{2}m_A\right)\varphi^A,$$
$$(\nabla_1 + i\nabla_2)\varphi^A = 0. \qquad (4.5.5)$$

These are our master equations. Once these equations are satisfied the energy of the BPS object is given by Eq. (4.5.4).

Let us discuss the central charges (the surface terms) of the string, domain wall and monopole in more detail. Say, in the string case, the three-dimensional integral in the first term in Eq. (4.5.4) gives the length of the string times its flux. In the wall case, the two-dimensional integral in the second term in (4.5.4) gives the area of the wall times its tension. Finally, in the monopole case the integral in the last term in Eq. (4.5.4) gives the magnetic-field flux. This means that the first-order master equations (4.5.5) can be used to study *strings, domain walls, monopoles and all their possible junctions.*

It is instructive to check that the wall, the string and the monopole solutions, separately, satisfy these equations. For the domain wall this check was done in [37] where we used these equations to study the string-wall junctions (we review this in Chapter 9). Let us consider the string solution. Then the scalar fields a and a^a are given by their VEV's. The gauge flux is directed along the z axis, so that $F_1^* = F_2^* = F_1^{*a} = F_2^{*a} = 0$. All fields depend only on the perpendicular coordinates x_1 and x_2. As a result, the first two equations and the fifth one in (4.5.5) are trivially satisfied. The third and the fourth equations reduce to the first two equations in Eq. (4.2.10). The last equation in (4.5.5) reduces to the last equation in (4.2.10).

Now, turn to the monopole solution. The 't Hooft–Polyakov monopole equations [105, 106] arise from those in Eq. (4.5.5) in the limit $\xi = 0$. Then all quark fields vanish, and Eq. (4.5.5) reduces to the standard first-order equations for the BPS 't Hooft–Polyakov monopole (see Section 3.4),

$$F_k^{*a} + \sqrt{2}\, D_k\, a^a = 0. \tag{4.5.6}$$

The U(1) scalar field a is given by its VEV while the U(1) gauge field vanishes.

Now, Eq. (4.5.4) shows that the central charge of the SU(2) monopole is determined by $\langle a^a \rangle$ which is proportional to the quark mass difference, see (4.1.11). Thus, for the monopole on the Coulomb branch (i.e. at $\xi = 0$) Eq. (4.5.4) yields

$$M_M = \frac{4\pi (m_{\ell_0+1} - m_{\ell_0})}{g_2^2}. \tag{4.5.7}$$

This coincides, of course, with the Seiberg–Witten result [2] in the weak coupling limit. As we will see shortly, the same expression continues to hold even if $\Delta m_{AB} \ll \sqrt{\xi}$ (provided that Δm_{AB} is still much larger than $\Lambda_{\mathrm{SU}(N)}$). An explanation will be given in Section 4.6.

The Abelian version of the first-order equations (4.5.5) were derived in Ref. [142] where they were exploited to find the 1/4 BPS-saturated solution for the wall-string junction. The non-Abelian equations (4.5.5) in the SU(2)$\times$ U(1) theory were derived in [165] where the confined monopoles as string junctions were considered at $\Delta m \neq 0$. Then the non-Abelian equations (4.5.5) were extensively used in the analysis [37] of the wall-string junctions in the problem of non-Abelian strings ending on a stack of domain walls. Next, Eqs. (4.5.5) for the confined monopoles as string junctions were solved in [132] in the SU(2) $\times$ U(1) theory. Below we will review this solution. Later all 1/4 BPS solutions for junctions (in particular, semilocal string junctions) were found in [172].

4.5.2 The string junction solution in the quasiclassical regime

Now we will apply our master equations at $N = 2$ in order to find the junction of the $S^a = (0, 0, 1)$ and $S^a = (0, 0, -1)$-strings via an SU(2) monopole in the quasiclassical limit. We assume that the $S^a = (0, 0, 1)$-string is at negative z, while the $S^a = (0, 0, -1)$-string is at positive z. We will show that the solution of the BPS equations (4.5.5) of the four-dimensional bulk theory is determined by the kink solution in the two-dimensional sigma model (4.4.48).

To this end we will look for the solution of equations (4.5.5) in the following *ansatz*. Assume that the solution for the string junction is given, to the leading order in $\Delta m/\sqrt{\xi}$, by the same string configuration (4.4.20), (4.4.5) and (4.4.42) which we dealt with previously (in the case $\Delta m \neq 0$) with S^a slowly varying functions of z, to be determined below, replacing the constant moduli vector S^a.

Now the functions $S^a(z)$ satisfy the boundary condition

$$S^a(-\infty) = (0, 0, 1), \tag{4.5.8}$$

while

$$S^a(\infty) = (0, 0, -1). \tag{4.5.9}$$

This *ansatz* corresponds to the non-Abelian string in which the vector S^a slowly rotates from (4.5.8) at $z \to -\infty$ to (4.5.9) at $z \to \infty$. We will show that the representation (4.4.20), (4.4.5) and (4.4.42) solves the master equations (4.5.5) provided the functions $S^a(z)$ are chosen in a special way.

Note that the first equation in (4.5.5) is trivially satisfied because the field a is constant and $F_1^* = F_2^* = 0$. The last equation reduces to the first two equations in (4.2.11) because it does not contain derivatives with respect to z and, therefore, is satisfied for arbitrary functions $S^a(z)$. The same remark applies also to the third equation in Eq. (4.5.5), which reduces to the third equation in (4.2.11).

Let us inspect the fifth equation in Eq. (4.5.5). Substituting our *ansatz* in this equation and using the formula (4.4.13) for ρ we find that this equation is satisfied provided $S^a(z)$ are chosen to be the solutions of the equation

$$\partial_3 S^a = \Delta m (\delta^{a3} - S^a S^3). \tag{4.5.10}$$

Below we will show that these equations are nothing but the first-order kink equations in the massive CP(1) model.

By the same token, we can consider the second equation in (4.5.5). Upon substituting there our *ansatz*, it reduces to Eq. (4.5.10) too. Finally, consider the fourth equation in (4.5.5). One can see that in fact it contains an expansion in the parameter $\Delta m^2/\xi$. This means that the solution we have just built is not exact; it has corrections of the order of $O(\Delta m^2/\xi)$. To the leading order in this parameter the fourth equation in (4.5.5) reduces to the last equation in (4.2.11). In principle, one could go beyond the leading order. Solving the fourth equation in (4.5.5) in the next-to-leading order would allow one to determine $O(\Delta m^2/\xi)$ corrections to our solution.

Let us dwell on the meaning of Eq. (4.5.10). This equation is nothing but the equation for the kink in the CP(1) model (4.4.48). To see this let us write the

Bogomol'nyi representation for kinks in the model (4.4.48). The energy functional can be rewritten as

$$E = \frac{\beta}{2} \int dz \left\{ \left| \partial_z S^a - \Delta m \left(\delta^{a3} - S^a S^3 \right) \right|^2 + 2 \Delta m \, \partial_z S^3 \right\}. \quad (4.5.11)$$

The above representation implies the first-order equation (4.5.10) for the BPS-saturated kink. It also yields $2\beta \Delta m$ for the kink mass.

Thus, we have demonstrated that the solution describing the junction of the $S^a = (0, 0, 1)$ and $S^a = (0, 0, -1)$ Z_2 strings is given by the non-Abelian string with a slowly varying orientation vector S^a. The variation of S^a is described in terms of the kink solution of the $(1 + 1)$-dimensional CP(1) model with the twisted mass.

In conclusion, we would like to match the masses of the four-dimensional monopole and two-dimensional kink. The string mass and that of the string junction is given by the first and the last terms in the surface energy (4.5.4) (the second term vanishes). The first term obviously reduces to

$$M_{\text{string}} = 2 \pi \xi L, \quad (4.5.12)$$

i.e. proportional to the total string length L. Note that both the $S^a = (0, 0, 1)$ and $S^a = (0, 0, -1)$ strings have the same tension (4.2.12). The third term should give the mass of the the monopole. The surface integral in this term reduces to the flux of the $S^a = (0, 0, -1)$-string at $z \to \infty$ minus the flux of the $S^a = (0, 0, 1)$-string at $z \to -\infty$. The F^{*3} flux of the $S^a = (0, 0, -1)$-string is 2π while the F^{*3} flux of the $S^a = (0, 0, 1)$-string is -2π. Thus, taking into account Eq. (4.1.11), we get

$$M_M = \frac{4\pi}{g_2^2} \Delta m. \quad (4.5.13)$$

Note, that although we discuss the monopole in the confinement phase at $|\Delta m| \ll \sqrt{\xi}$ (in this phase it is a junction of two strings), nevertheless the Δm and g_2^2 dependence of its mass coincides with the result (4.5.7) for the unconfined monopole on the Coulomb branch (i.e. at $\xi = 0$). This is no accident – there is a deep theoretical reason explaining the validity of this unified formula. A change occurs only in passing to a highly quantum regime depicted in the right lower corner of Fig. 4.3. We will discuss this regime shortly in Section 4.5.3.

It is instructive to compare Eq. (4.5.13) with the kink mass in the effective CP(1) model on the string world sheet. As was mentioned, the surface term in Eq. (4.5.11) gives

$$M_{\text{kink}} = 2 \beta \Delta m. \quad (4.5.14)$$

Now, expressing the two-dimensional coupling constant β in terms of the coupling constant of the microscopic theory, see Eq. (4.4.15), we obtain

$$M_{\text{kink}} = \frac{4\pi}{g_2^2}\Delta m, \qquad (4.5.15)$$

thus verifying that the four-dimensional calculation of M_M and the two-dimensional calculation of M_{kink} yield the same,

$$M_M = M_{\text{kink}}. \qquad (4.5.16)$$

Needless to say, this is in full accordance with the physical picture that emerged from our analysis, that the two-dimensional CP(1) model is nothing but the macroscopic description of the confined monopoles occurring in the four-dimensional microscopic Yang–Mills theory. Technically the coincidence of the monopole and kink masses is based on the fact that the integral in the definition (4.4.10) of the sigma-model coupling β reduces to unity.

4.5.3 The strong coupling limit

Here we will consider the limit of small Δm_{AB}, when the effective world sheet theory develops a strong coupling regime. For illustrative purposes we will consider the simplest case, $N = 2$. Generalization to generic N is straightforward.

As we further diminish $|\Delta m|$ approaching Λ_σ and then send Δm to zero we restore the global SU(2)$_{C+F}$ symmetry. In particular, on the Coulomb branch, the SU(2) × U(1) gauge symmetry is restored. In this limit the monopole size grows, and, classically, it would explode. Moreover, the classical formula (4.5.13) interpreted literally shows that the monopole mass vanishes (see the discussion of the so-called "monopole clouds" in [112] for a review of the long-standing issue of understanding what becomes of the monopoles upon restoration of the non-Abelian gauge symmetry). Thus, classically one would say that the monopoles disappear.

That's where quantum effects on the confining string take over. As we will explain below, they materialize the confined non-Abelian monopole as a well-defined stable object [132].

While the string thickness (in the transverse direction) is $\sim \xi^{-1/2}$, the z-direction size of the kink representing the confined monopole in the highly quantum regime is much larger, $\sim \Lambda_\sigma^{-1}$, see the lower right corner in Fig. 4.3. Still, it remains finite in the limit $\Delta m \rightarrow 0$, stabilized by non-perturbative effects in the world sheet CP(1) model. This is due to the fact that the CP(N − 1) models develop a mass gap, and no massless states are present in the spectrum, see Section 4.4.3. Moreover, the mass of the confined monopole (the CP(1) model kink) is also determined by the scale Λ_σ. This sets the notion of what the confined non-Abelian monopole is. It is a kink in the massless two-dimensional CP(1) model [132].

We can get a more quantitative insight into the physics of the world-sheet theory at strong coupling if we invoke the exact BPS spectrum of the twisted-mass-deformed CP($N-1$) model obtained in [30]. For a detailed discussion in CP(1) see Section 3.5.

The exact expression for the central charge in CP($N-1$) with twisted mass was derived [30] by generalizing Witten's analysis [157] that had been carried out previously for the massless case. The BPS states saturate the central charge Z defined in Eq. (2.3.4). The exact formula for this central charge is

$$Z_{2d} = i\,\Delta m\, q + m_D\, T, \qquad (4.5.17)$$

where the subscript $2d$ reminds us that the model in question is two-dimensional. The subscript D in m_D appears for historical reasons, in parallel with the Seiberg–Witten solution (it stands for dual). Furthermore, T is the topological charge of the kink under consideration, $T = \pm 1$, while the parameter q

$$q = 0,\ \pm 1,\ \pm 2, \ldots \qquad (4.5.18)$$

This global U(1) charge of the "dyonic" states arises due to the presence of a U(1) group unbroken in (4.4.48) by the twisted mass (the SU(2)$_{C+F}$ symmetry is broken down to U(1) by $\Delta m \neq 0$).

The quantity m_D was introduced [30] in analogy with a_D of Ref. [2]. In the case $N = 2$ it has the form presented in Eq. (3.5.27) with the substitutions

$$m \to \Delta m, \qquad \Lambda \to \Lambda_\sigma, \qquad (4.5.19)$$

where Δm is now assumed to be complex. The two-dimensional central charge is normalized in such a way that $M_{\text{kink}} = |Z_{2d}|$.

As we discussed in Section 3.5, there are no massless states in the CP(1) model at $\Delta m = 0$. In particular, the kink (confined monopole) mass is

$$M_M = \frac{2}{\pi}\Lambda_\sigma, \qquad (4.5.20)$$

as it is clear from (3.5.27). On the other hand, in this limit both the last term in (4.5.4) and the surface term in (4.5.11) vanish for the monopole and kink masses, respectively. What's wrong?

This puzzle was solved by the following observation: anomalous terms in the central charges of both four-dimensional and two-dimensional SUSY algebras are present in these theories. In two dimensions the anomalous terms were obtained in [33, 34]. In four dimensions the bifermion anomalous term was discovered in [132]. We refer the reader to Section 3.4.2 for a more detailed discussion.

In the bulk theory the central charge associated with the monopole is defined through the anticommutator

$$\{\bar{Q}_{\dot\alpha}^f \; \bar{Q}_{\dot\beta}^g\} = 2\,\varepsilon_{\dot\alpha\dot\beta}\,\varepsilon^{fg}\,\bar{Z}_{4d}, \qquad (4.5.21)$$

where $\bar{Z}_{4d}$ is an SU(2)$_R$ singlet; the subscript $4d$ will remind us of four dimensions. It is most convenient to write $\bar{Z}_{4d}$ as a topological charge (i.e. the integral over a topological density),

$$\bar{Z}_{4d} = \int d^3x\, \bar{\zeta}^0(x). \qquad (4.5.22)$$

In the model at hand[10]

$$\bar{\zeta}^\mu = \frac{1}{\sqrt{2}}\varepsilon^{\mu\nu\rho\sigma}\,\partial_\nu\left(\frac{i}{g_2^2}\,a^a\,F_{\rho\sigma}^a + \frac{i}{g_1^2}\,a\,F_{\rho\sigma} - \frac{i}{2\pi^2}\,a^a\,F_{\rho\sigma}^a\right.$$

$$\left. + \frac{i}{8\sqrt{2}\pi^2}\left[\lambda_{f\alpha}^a(\sigma_\rho)^{\alpha\dot\alpha}(\bar\sigma_\sigma)_{\dot\alpha\beta}\lambda^{af\beta} + 2g_2^2\tilde\psi_{A\alpha}(\sigma_\rho)^{\alpha\dot\alpha}(\bar\sigma_\sigma)_{\dot\alpha\beta}\psi^{A\beta}\right]\right).$$

$$(4.5.23)$$

Note that the general structure of the operator in the square brackets is unambiguously fixed by dimensional arguments, the Lorentz symmetry and other symmetries of the bulk theory. The numerical coefficient was first found in [132] by matching the monopole and kink masses at $\Delta m = 0$.

The above expression is an operator equality. In the low-energy limit, the Seiberg–Witten exact solution allows one to obtain the full matrix element of the operator on the right-hand side (which includes all perturbative and non-perturbative corrections) by replacing a by a_D.

The fermion part of the anomalous term plays a crucial role in the Higgs phase for the confined monopole. On the Coulomb branch it does not contribute to the mass of the monopole due to a fast fall off of the fermion fields at infinity. On the Coulomb branch the bosonic anomalous terms become important. The relationship between the 't Hooft–Polyakov monopole mass and the $\mathcal{N} = 2$ central charge is analyzed in [38], which identifies an anomaly in the central charge explaining a constant (i.e. non-logarithmic) term in the monopole mass on the Coulomb branch. The result of Ref. [38] is in agreement with the Seiberg–Witten formula for the monopole mass. In Section 3.4.2 we presented the operator form of the central charge anomaly.

---

[10] In Eq. (4.5.23) in the bosonic part we keep only terms containing the magnetic field $\vec{B}$ and drop those with the electric field $\vec{E}$ which are relevant for dyons. For more details see Section 3.4.2.

Note, that the coefficient in front of the fermionic term involving λ-fermions in (4.5.23) coincides with the one in (3.4.20) obtained by supersymmetrization of the bosonic anomalous term.

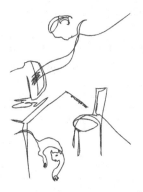

4.6 Two-dimensional kink and four-dimensional Seiberg–Witten solution

Why is the 't Hooft–Polyakov monopole mass (i.e. that on the Coulomb branch at $\xi = 0$) given by the same formula (4.5.7) as the mass (4.5.13) of the strongly confined large-ξ monopole (subject to condition $\sqrt{\xi} \gg \Delta m$)?

This fact was noted in Section 4.5.2. Now we will explain the reason lying behind this observation [132, 133]. En route, we will explain another striking observation made in Ref. [30]. A remarkably close parallel between four-dimensional SU(2) Yang–Mills theory with $N_f = 2$ and the two-dimensional CP(1) model was noted, at an observational level, by virtue of comparison of the corresponding central charges. The observation was made on the Coulomb branch of the Seiberg–Witten theory, with unconfined monopoles/dyons of the 't Hooft–Polyakov type. Valuable as it is, the parallel was quite puzzling since the solution of the CP(1) model seemed to have no physics connection to the Seiberg–Witten solution. The latter gives the mass of the unconfined monopole in the Coulomb regime at $\xi = 0$ while the CP(1) model emerges only in the Higgs regime of the bulk theory.

We want to show that the reason for the correspondence mentioned above is that in the BPS sector (and *only* in this sector) the parameter ξ, in fact, does not appear in relevant formulae. Therefore, one can vary ξ at will, in particular, making it less than $|\Delta m|$ or even tending to zero, where CP(1) is no more the string world sheet theory for our bulk model. Nevertheless, the parallel expressions for the central charges and other BPS data in four dimensions and two dimensions, trivially established at $|\Delta m| \ll \xi$, will continue to hold even on the Coulomb branch. The "strange coincidence" we observed in Section 4.5.2 is no accident. We deal with an exact relation which stays valid including both perturbative and non-perturbative corrections.

Physically the monopole in the Coulomb phase is very different from the one in the confinement phase, see Fig. 4.3. In the Coulomb phase it is a 't Hooft–Polyakov monopole, while in the confinement phase it becomes related to a junction of two non-Abelian strings. Still let us show that the masses of these two objects are given by the same expression,

$$M_M^{\text{Coulomb}} = M_M^{\text{confinement}} \tag{4.6.1}$$

provided that Δm and the gauge couplings are kept fixed. The superscripts refer to the Coulomb and monopole-confining phases, respectively.

The crucial observation here is that the mass of the monopole cannot depend on the FI parameter ξ. Start from the monopole in the Coulomb phase at $\xi = 0$. Its mass is given by the exact Seiberg–Witten formula [3]

$$M_M^{\text{Coulomb}} = \sqrt{2} \left| a_D^3 \left(a^3 = -\frac{\Delta m}{\sqrt{2}} \right) \right|$$

$$= \left| \frac{\Delta m}{\pi} \ln \frac{\Delta m}{\Lambda_{\text{SU}(2)}} + \Delta m \sum_{k=0}^{\infty} c_k \left(\frac{\Lambda}{\Delta m} \right)^{2k} \right|, \tag{4.6.2}$$

where a_D^3 is the dual Seiberg–Witten potential for the SU(2) gauge group. We take into account the fact that for $N_f = 2$ the first coefficient of the β function is 2.

In Eq. (4.6.2) $a^3 = -\Delta m/\sqrt{2}$ is the argument of a_D^3, the logarithmic term takes into account the one-loop result (4.1.19) for the SU(2) gauge coupling at the scale Δm, while the power series represents instanton-induced terms – small corrections at large a.

Now, if we switch on a small FI parameter $\xi \neq 0$ in the theory, on dimensional grounds we could expect corrections to the monopole mass in powers of $\sqrt{\xi}/\Lambda_{\text{SU}(2)}$ and/or $\sqrt{\xi}/\Delta m$ in Eq. (4.6.2).

But ... these corrections are *forbidden* by the U(1)$_R$ charges. Namely, the U(1)$_R$ charges of $\Lambda_{\text{SU}(2)}$ and Δm are equal to 2 (and so is the U(1)$_R$ charge of the central charge under consideration) while ξ has a vanishing U(1)$_R$ charge. For convenience, the U(1)$_R$ charges of different fields and parameters of the microscopic theory are collected in Table 4.1. Thus, neither $(\sqrt{\xi}/\Lambda_{\text{SU}(2)})^k$ nor $(\sqrt{\xi}/\Delta m)^k$ can appear.

By the same token, we could start from the confined monopole at large ξ, and study the dependence of the monopole (string junction) mass as a function of ξ as we reduce ξ. Again, the above arguments based on the U(1)$_R$ charges tell us that

Table 4.1. *The* U(1)$_R$ *charges of fields and parameters of the bulk theory.*

Field/parameter	a	a^a	λ^α	q	ψ^α	m_A	$\Lambda_{SU(N)}$	ξ
U(1)$_R$ charge	2	2	1	0	-1	2	2	0

no corrections in powers of $\Lambda_{SU(2)}/\sqrt{\xi}$ and $\Delta m/\sqrt{\xi}$ can appear. This leads us to Eq. (4.6.1).

Another way leading to the same conclusion is to observe that the monopole mass depends on a (anti)holomorphically, cf. Seiberg–Witten's formula (4.6.2). Thus, it cannot depend on the FI parameter ξ which is not holomorphic (it is a component of the SU(2)$_R$ triplet [127, 35]).

Now let us turn to the fact that the mass of the monopole in the confinement phase is given by the kink mass in the CP(1) model, see (4.5.16). In this way we obtain

$$M_M^{\text{Coulomb}} \leftrightarrow M_M^{\text{confinement}} \leftrightarrow M_{\text{kink}}. \qquad (4.6.3)$$

In particular, at one loop, the kink mass is determined by renormalization of the CP(1)-model coupling constant β, while the monopole mass on the Coulomb branch is determined by the renormalization of g^2. This leads to the relation

$$\Lambda_\sigma = \Lambda_{SU(2)}$$

between the two- and four-dimensional dynamical scales. It was noted earlier as a "strange coincidence," see Eq. (4.4.35). The first coefficient of the β functions is two (N for generic N) for both theories. Now we know the physical reason behind this coincidence.

Clearly, the above relation can be generalized (cf. [30, 133]) to cover the SU(N)×U(1) case with $N_f = N$ flavors on the four-dimensional side, and CP($N - 1$) sigma models on the two-dimensional side.

This correspondence can be seen in more quantitative terms [30, 133]. Four-dimensional U(N) SQCD with $\mathcal{N} = 2$ and $N_f = N$ flavors is described by the degenerate Seiberg–Witten curve

$$y^2 = \frac{1}{4}\left[\prod_{i=1}^{N}(x + \tilde{m}_i) - \Lambda_{SU(N)}^N\right]^2 \qquad (4.6.4)$$

in the special point (4.1.11) on the Coulomb branch which becomes a quark vacuum upon the ξ deformation. The periods of this curve give the BPS spectrum of the two-dimensional CP($N-1$) model [30]. We quoted this spectrum for CP(1) in Eqs. (4.5.17) and (3.5.27).

In fact, Dorey demonstrated [30] that the BPS spectra of the two-dimensional CP($N-1$) model and four-dimensional SU(N) SQCD coincide with each other if one chooses a point on the Coulomb branch corresponding to the baryonic Higgs branch defined by the condition $\sum m_A = 0$ (in the SU(2) case the gauge equivalent choice is to set $m_1 = m_2$).

At the same time, we observe that the BPS spectra of the *massive* states in the SU(2) and U(2) theories, respectively, coincide in the corresponding quark vacua upon identification of m_A of the SU(N) theory with $\tilde{m}_A$ of the U(N) theory. In particular, in the $N = 2$ case one must identify $m_1 = m_2$ of the SU(2) theory with $\Delta m/2$ of the U(2) theory. Note that the vacuum (4.1.11) and (4.1.14) of the U(N) theory is an isolated vacuum rather than a root of a Higgs branch. There are no massless states in the U(N) bulk theory in this vacuum, see Section 4.1.2 for more details.

Note also that the BPS spectra of both theories include not only the monopole/kink and "dyonic" states but elementary excitations with $T = 0$ as well. On the two-dimensional side they correspond to elementary fields n^ℓ in the large Δm_{AB} limit. On the four-dimensional side they correspond to non-topological (i.e. $T = 0$ and $q = \pm 1$) BPS excitations of the string with masses proportional to Δm_{AB} confined on the string.

The latter can be interpreted as follows. Inside the string the squark profiles vanish, effectively bringing us into the Coulomb branch ($\xi = 0$) where the W bosons and quarks would become BPS-saturated states in the bulk. Say, for $N = 2$ on the Coulomb branch, the W boson and off-diagonal quark mass would reduce to Δm. Hence, the $T = 0$ BPS excitation of the string is a wave of such W bosons/quarks propagating along the string. One could term it a "confined W boson/quark." It is localized in the perpendicular but not in the longitudinal direction. What is important, it has no connection with the bulk Higgs phase W bosons, which are non-BPS and are much heavier than Δm. Nor do these non-topological excitations have connection to the bulk quarks of our bulk model, which are not BPS-saturated too.

To conclude, let us mention that Tong compared [173] a conformal theory with massless quarks and monopoles arising on the Coulomb branch of the four-dimensional $\mathcal{N} = 2$ SQCD (upon a special choice of the mass parameters Δm_{AB}), at the so-called Argyres–Douglas point [174], with the twisted-mass-deformed two-dimensional CP($N-1$) model.

The coincidence of the monopole and kink masses explained above ensures that the CP($N-1$) model flows to a non-trivial conformal point at these values of Δm_{AB}. The scaling dimensions of the chiral primary operators in four- and two-dimensional conformal theories were shown to agree [173]; a very nice result, indeed.

4.7 More quark flavors

In this section we will abandon the assumption $N_F = N$ and consider the theory (4.1.7) with more fundamental flavors, $N_F > N$. In this case we have a number of isolated vacua such as (4.1.11) and (4.1.14), in which N squarks out of N_f develop VEV's, while the adjoint VEV's are determined by the mass terms of these quarks, as in Eq. (4.1.11).

Now, let us focus on the equal mass case. Then the isolated vacua coalesce, and a Higgs branch develops from the common root whose location on the Coulomb branch is given by Eq. (4.1.11) (with all masses set equal). The dimension of this branch is $4N(N_f - N)$, see [143, 140]. The Higgs branch is noncompact and has a hyper-Kähler geometry [3, 143]. It has a compact base manifold defined by the condition

$$\bar{\tilde{q}}^{kA} = q^{kA}. \tag{4.7.1}$$

The dimension of this manifold is twice less than the total dimension of the Higgs branch, $2N(N_f - N)$, which implies 4 for $N_f = 3$ and 8 for $N_f = 4$ in the simplest $N = 2$ case. The BPS string solutions exist only on the base manifold of the Higgs branch. The flux tubes become non-BPS-saturated if we move away from the base along noncompact directions [175]. Therefore, we will limit ourselves to the vacua which belong to the base manifold.

Strings that emerge in multiflavor theories, i.e. $N_f > N$ (typically on the Higgs branches), as a rule are not conventional ANO strings. Rather, they become the so-called semilocal strings (for a comprehensive review see [147]). The simplest model where the semilocal strings appear is the Abelian Higgs model with two complex flavors

$$S_{AH} = \int d^4x \left\{ \frac{1}{4g^2} F_{\mu\nu}^2 + |\nabla_\mu q^A|^2 + \frac{g^2}{8} (|q^A|^2 - \xi)^2 \right\}, \qquad (4.7.2)$$

where $A = 1, 2$ is the flavor index.

If $\xi \neq 0$ the scalar fields develop VEV's breaking the U(1) gauge group. The photon field becomes massive, together with one real scalar field.

In fact, for the particular choice of the quartic coupling made in Eq. (4.7.2) this scalar field has the same mass as the photon. In fact, the model (4.7.2) is the bosonic part of a supersymmetric theory; the flux tubes are classically BPS-saturated. The topological reason for the existence of the ANO flux tubes is that

$$\pi_1[U(1)] = Z$$

for the U(1) gauge group. On the other hand, in Eq. (4.7.2) we can pass to the low-energy limit integrating out the massive photon and its scalar counterpart. This will lead us to a four-dimensional sigma model on the manifold

$$|q^A|^2 = \xi. \qquad (4.7.3)$$

The vacuum manifold (4.7.3) has dimension $4 - 1 - 1 = 2$, where we subtract one real condition mentioned above, as well as one phase that can be gauged away. Thus, the manifold (4.7.3) represents a two-dimensional sphere S_2. The low-energy limit of the theory (4.7.2) is the O(3) sigma model.

We should remember that

$$\pi_2[S_2] = \pi_1[U(1)] = Z,$$

and this is the topological reason for the existence of instantons in the two-dimensional O(3) sigma model [176]. Uplifted in four dimensions, these instantons become string-like objects (lumps).

Just as the O(3) sigma-model instantons, the semilocal strings possess two additional zero modes associated with its complexified size modulus ρ in the model (4.7.2). Hence, the semilocal strings interpolate between the ANO strings and two-dimensional sigma-model instantons uplifted in four dimensions. At $\rho = 0$ we have the ANO string while at $\rho \to \infty$ the string becomes nothing but the two-dimensional instanton elevated in four dimensions. At generic $\rho \neq 0$ the semilocal string is characterized by a power fall-off of the profile functions at infinity, to be contrasted with the exponential fall-off characteristic of the ANO string.

Now, if we return to our non-Abelian theory (4.1.7), we will see that the semilocal strings in this theory have size moduli, in addition to the $2(N-1)$ orientational moduli n^ℓ. The total dimension of the moduli space of the semilocal string was shown [130] to be

$$2N_f = 2 + 2(N-1) + 2(N_f - N), \qquad (4.7.4)$$

where the first, the second and the third terms above correspond to the translational, orientational and the size moduli.

No studies of geometry of the moduli space of the semilocal strings were carried out for quite some time due to infrared problems. It was known [177, 178] that the size-zero modes are logarithmically non-normalizable in the infrared, as is the case for the sigma-model instantons in two dimensions. This problem was addressed in [179] where non-Abelian strings in the U(2) gauge theory were treated. The effective theory on the string world sheet was shown to have the form

$$S^{(1+1)} = \beta\, M_W \int dt\, dz \left\{ \frac{\rho^2}{4} (\partial_k S^a)^2 + |\partial_k \rho_i|^2 \right\} \ln \frac{1}{|\rho|\,\delta m}, \qquad (4.7.5)$$

where M_W is the W-boson mass, see Eq. (4.1.16). The subscript $i = 3, \ldots, N_f$, while ρ_i stand for $(N_f - 2)$ complex fields associated with the size moduli. The parameter δm here measures small quark mass differences, acting as an infrared regulator. It is necessary to introduce this infrared parameter, slightly lifting the size moduli ρ_i, in order to regularize the infrared logarithmic divergence.

The metric (4.7.5) is derived in [179] for large – but not too large – values of $|\rho|^2 \equiv |\rho_i^2|$ lying inside the window

$$\frac{1}{M_W} \ll |\rho| \ll \frac{1}{\delta m}. \qquad (4.7.6)$$

The inequality on the left-hand side refers to the limit in which the semilocal string becomes an O(3) sigma-model lump. The inequality on the right-hand side ensures the validity of the logarithmic approximation. The action (4.7.5) was obtained in the logarithmic approximation.

For ρ_i's lying inside the window (4.7.6), with a logarithmic accuracy, one can introduce new variables

$$z_i = \rho_i \left[M_W^2 \ln \frac{1}{|\rho|\,\delta m} \right]^{1/2}. \qquad (4.7.7)$$

In terms of these new variables the metric of the world sheet theory (4.7.5) was shown[11] to become flat [179]. Corrections to this flat metric run in powers of

$$\frac{1}{M_W |\rho|} \quad \text{and} \quad \left(\ln \frac{1}{|\rho| \, \delta m} \right)^{-1}.$$

These corrections have not yet been calculated within the field-theory approach.

On the other hand, the very same problem was analyzed from the D-brane theory side. Using brane-based arguments Hanany and Tong conjectured [130, 133] (see also Ref. [152]) that the effective theory on the world sheet of the non-Abelian semilocal string is given by the strong-coupling limit ($e^2 \to \infty$) of the following two-dimensional gauge theory:

$$S = \int d^2 x \left\{ 2\beta \, |\nabla_k n^\ell|^2 + 2\beta \, |\nabla_k z_i|^2 + \frac{1}{4e^2} F_{kl}^2 + \frac{1}{e^2} |\partial_k \sigma|^2 \right.$$

$$+ 4\beta \left| \sigma - \frac{\tilde{m}_\ell}{\sqrt{2}} \right|^2 |n^\ell|^2 + 4\beta \left| \sigma - \frac{\tilde{m}_i}{\sqrt{2}} \right|^2 |z_i|^2$$

$$\left. + 2e^2 \, \beta^2 \left[|n^\ell|^2 - |z_i|^2 - 1 \right]^2 \right\}, \tag{4.7.8}$$

where $\ell = 1, \ldots, N$ and $i = N + 1, \ldots, N_f$. Furthermore, z_i denote $(N_f - N)$ complex fields associated with the size moduli. The fields n^ℓ and z_i have the charges $+1$ and -1 with respect to the U(1) gauge field in Eq. (4.7.8). This theory is similar to the model (4.4.51) describing the $N_f = N$ non-Abelian strings.

The Hanany–Tong conjecture is supported by yet another argument. As was discussed in Section 4.6, the BPS spectrum of dyons on the Coulomb branch of the four-dimensional theory must coincide with the BPS spectrum in the two-dimensional theory on the string world sheet. We expect that this correspondence extends to theories with $N_f > N$. The two-dimensional theory (4.7.8) was studied in [181] where it was shown that its BPS spectrum agrees with the spectrum of four-dimensional U(N) SQCD with N_f flavors. In particular, the one-loop coefficient of the β function is $2N - N_f$ in both theories. This leads to the identification of their scales, see Eq. (4.4.35). As a matter of fact, Ref. [181] deals with the SU(N) theory at the root of the baryonic Higgs branch, much in the same vein as [30]. However, as was explained in Section 4.6, the BPS spectra of the massive states in these four-dimensional theories are the same.

---

[11] Warning: a different metric on the moduli space of the non-Abelian semilocal string was suggested in [180]. It has a kinetic cross-term for the orientational and size moduli fields. However, at large ρ_i, inside the allowed window (4.7.6), this metric is also flat.

The above argument shows that the two-dimensional theory (4.7.8) is a promising candidate for an effective theory on the semilocal string world sheet. In particular, the metric in (4.7.8) is asymptotically flat. The variables z_i in (4.7.8) should be identified with the ones in Eq. (4.7.7) introduced within the field-theory framework in Ref. [179]. It is quite plausible that corrections to the flat metric in powers $1/(M_W |\rho|)$ are properly reproduced by the world sheet theory (4.7.8). Nevertheless, the results of [179] clearly demonstrate the approximate nature of the world sheet theory (4.7.8). Namely, corrections at large ρ_i suppressed by large infrared logarithms $(\ln (1/|\rho| \, \delta m))^{-1}$ are certainly not captured in Eq. (4.7.8).

The implication of the "semilocal nature" of the semilocal strings which is most important from the physical standpoint is the loss of the monopole confinement [175, 179] i.e. the loss of the Meissner effect. To study the monopole confinement as a result of the squark condensation we must consider a string of a finite length L stretched between a heavy probe monopole and antimonopole from the $SU(N + 1)/SU(N) \times U(1)$ sector. The ANO string has a typical transverse size $(g\sqrt{\xi})^{-1}$. If L is much larger than this size the energy of this probe configuration is

$$V(L) = TL, \qquad\qquad (4.7.9)$$

where T is the string tension. The linear potential in Eq. (4.7.9) ensures confinement of monopoles.

For semilocal strings this conventional picture drastically changes. Now the transverse string size can be arbitrarily large. Imagine a configuration in which the string transverse size becomes much larger than L. Then we will clearly deal with the three-dimensional rather than two-dimensional problem. The monopole flux is no longer trapped in a narrow flux tube. Instead, it freely spreads over a large three-dimensional volume, of the size of order of L in all directions. Obviously, this will give rise to a Coulomb-type potential between the probe monopoles,

$$V(L) \sim 1/L, \qquad\qquad (4.7.10)$$

possibly augmented by logarithms of L. At large L the energy of this configuration is lower than the one of the flux-tube configuration (4.7.9); therefore, it is energetically favorable.

To summarize, semilocal strings can indefinitely increase their transverse size and effectively disintegrate, so that the linear potential (4.7.9) gives place to the Coulomb potential (4.7.10). In fact, lattice studies unambiguously show that the semilocal string thickness always increases upon small perturbations [182].

Formation of semilocal strings on the Higgs branches leads to a dramatic physical effect – deconfinement.

4.8 Non-Abelian k-strings

In this section we will briefly review how multi-strings, with the winding number $k > 1$, can be constructed. One can consider them as bound states of k BPS elementary strings. The Bogomol'nyi representation (4.2.9) implies that the tension of the BPS-saturated k-string is determined by its total U(1) flux, $2\pi k$. This entails, in turn, that in $\mathcal{N} = 2$ SQCD, see Eq. (4.1.7), the k-string tension has the form (4.1.7)

$$T_k = 2\pi k \, \xi. \tag{4.8.1}$$

Equation (4.8.1) implies that the elementary strings that form composite k-strings do not interact.

If one considers k elementary strings, forming the given k-string, at large separations the corresponding moduli space obviously factorizes into k copies of the moduli spaces of the elementary strings. This suggests that the dimension of the total moduli space is

$$2kN_f = 2k + 2k(N-2) + 2k(N_f - N), \tag{4.8.2}$$

see (4.7.4). The total dimension is written as a sum of dimensions of the translational, orientational and size moduli spaces. This result was confirmed by the Hanany–Tong index theorem [130] which implies (4.8.2) at any separations. The moduli space of well-separated elementary strings forming the given k-string, say, at $N_f = N$ is

$$\frac{\left[C \times \mathrm{CP}(N-1) \right]^k}{S_k}, \tag{4.8.3}$$

where S_k stands for permutations of the elementary string positions.

An explicit solution for a non-Abelian 2-string at zero separation in the simplest bulk theory with $N = N_f = 2$ was constructed in [183]. It has a peculiar feature. If the orientation vectors of the two strings S_1^a and S_2^a are opposite, the composite 2-string becomes an Abelian ANO string. It carries no non-Abelian flux. Therefore, $SU(2)_{C+F}$ rotations act trivially on this particular string. This means that the internal moduli space of this string is singular [184, 183]. The section of the orientational moduli space corresponding to $S_1^a = -S_2^a$ degenerates into a point. In [183] it was argued that the internal moduli of the 2-string at zero separation is equivalent to $CP(2)/Z_2$. This differs by a discrete quotient from the result $CP(2)$ obtained in [184]. Later results obtained in [185, 186] confirm the $CP(2)/Z_2$ metric.

The metric on the k-string moduli space for generic k is not known. For Abelian k-strings exponential corrections to the flat metric were calculated in [187]. Exponentially small corrections are natural since in this case the vortices are characterized by an exponential fall off of their profile functions at large distances.

Hanany and Tong exploited [130, 133] a D-brane construction to obtain the k-string metric in terms of the Higgs branch of a two-dimensional gauge theory, see (4.4.51) and (4.7.8). What they came up with is an $\mathcal{N} = 2$ supersymmetric $U(k)$ gauge theory with N fundamental and $(N_f - N)$ anti-fundamental flavors n^ℓ and ρ_i, respectively, ($\ell = 1, \ldots, N$ and $i = N, \ldots, N_f$), plus an adjoint chiral multiplet Z. The D-term condition for this theory is

$$\frac{1}{2\beta}[\bar{Z}, Z] + n^\ell \bar{n}_\ell - \rho_i \bar{\rho}^i = 1. \tag{4.8.4}$$

The metric defined by this Higgs branch has corrections to the factorized metric which run in (inverse) powers of separations between the elementary strings. Thus, it exhibits a dramatic disagreement with the field-theory expectations. Still the metric is believed [130, 133, 13, 14] to correctly reproduce some data protected by supersymmetry, such as the BPS spectrum.

To derive all moduli of the general k-string solution the so-called moduli matrix method was developed in [188]. It was observed that the substitution

$$\varphi = S(z, \bar{z}) \, H_0(z), \quad A_1 + iA_2 = S^{-1} \bar{\partial}_z S, \tag{4.8.5}$$

solves the last of the first-order equations (4.2.10). Here $z = x_1 + ix_2$ and H_0 is an $N \times N_f$ matrix with a holomorphic dependence on z.

Then the equations for the gauge field strength in (4.2.10) yield an equation on $S(z, \bar{z})$ which is rather hard to solve in the general case. It was argued, however, that the factor S involves no new moduli parameters [188]. Therefore, all moduli parameters reside in the moduli matrix $H_0(z)$.

Determining $H_0(z)$ gives one a moduli space which agrees with the moduli space corresponding to the Higgs branch (4.8.4).

4.9 A physical picture of the monopole confinement

In this section we will return to our basic $\mathcal{N} = 2$ SQCD with the U(N) gauge group and $N_f = N$ flavors (4.1.7) and discuss an emerging physical picture of the monopole confinement. As was reviewed in detail in Section 4.5, elementary confined monopoles can be viewed as junctions of two elementary strings. Therefore, the physical spectrum of the theory includes monopole-antimonopole "mesons" formed by two elementary strings in a loop configuration shown in Fig. 4.4.

If spins of such "mesons" are of order one, their mass is of the order of the square root of the string tension $\sqrt{\xi}$. Deep in the quantum non-Abelian regime ($\tilde{m}^l = 0$), the CP($N - 1$)-model strings carry no average SU(N) magnetic flux [159],

$$\langle n^l \rangle = 0, \tag{4.9.1}$$

see Eq. (4.4.4). What they do carry is the U(1) magnetic flux which determines their tension.

Monopoles are seen in the world sheet theory as CP($N - 1$) kinks. At $\tilde{m}^l = 0$ they become non-Abelian too, much in the same way as strings. They carry no average SU(N) magnetic flux. (Unlike strings, even in the classical regime they do not carry the U(1) magnetic flux, see (4.5.1).)

Moreover, the monopoles acquire global flavor quantum numbers. We know that the CP($N - 1$) model kinks at strong coupling are described by the n^l fields [159, 120] and, therefore, in fact, they belong to the fundamental representation of the global SU(N)$_{C+F}$ group. This means that the monopole-antimonopole "mesons" formed by the string configuration shown in Fig. 4.4 can belong either to singlet or to

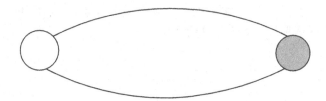

Figure 4.4. Monopole and antimonopole bound into a "meson." The binding is due to strings. Open and closed circles denote the monopole and antimonopole, respectively.

adjoint representations of the global "flavor" group $SU(N)_{C+F}$, in full accordance with our expectations.

Singlets resemble glueballs. In weakly coupled bulk theory ($g^2_{1,2} \ll 1$) the singlet mesons can decay into massive vector multiplets formed by gauge and quark fields, with mass (4.1.17), see Section 4.1.2. The monopole-antimonopole mesons with the adjoint flavor quantum numbers are also metastable in weakly coupled bulk theory, they decay into massive gauge/quark multiplets which carry the adjoint quantum numbers with respect to the global unbroken $SU(N)_{C+F}$ group and have masses determined by Eq. (4.1.16).

Two elementary strings of the monopole-antimonopole meson shown in Fig. 4.4 can form a non-BPS bound state. Hence, in practice the composite meson looks as if the monopole was connected to the antimonopole by a single string. In fact, there are indications that this is what happens in the theory at hand. Interactions of elementary Z_2 strings were studied in [140] in the simplest case $N = 2$. An interaction potential for the elementary Z_2 strings with $S^a = (0, 0, +1)$ and $S^a = (0, 0, -1)$ was found to be attractive at large distances,

$$U \sim - \left\{ M_{SU(2)} \, R \right\}^{-1/2} e^{-M_{SU(2)} R}, \qquad (4.9.2)$$

where R stands for the distance between two parallel strings. The gauge boson mass is given in Eq. (4.1.16). This attractive potential leads to formation of a bound state, a composite string.

Note that we have N distinct elementary strings. As was discussed in Section 4.4.3, in the quantum regime N elementary strings differ from each other by the value of the bifermion condensate of the CP($N - 1$) model fermions [156]. Therefore, the physical picture of the monopole confinement is not absolutely similar to what we expect in QCD, see the discussion in the beginning of this section. Namely, we have N different degenerate "mesons" (at $N > 2$) of the type discussed above, associated with N different elementary strings.

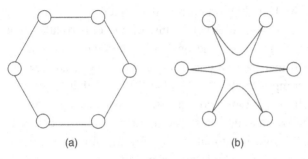

Figure 4.5. (a) A schematic picture of the "baryon" formed by monopoles and strings for $N = 6$; (b) The "baryon" acquires the shape of a star once the neighboring strings form non-BPS bound states.

In QCD (and in nature) we have instead a single meson with the given quantum numbers, plus its radial excitations which have higher masses. This is typical for BPS strings in supersymmetric gauge theories. We will see in Chapter 6 that in non-supersymmetric theories the situation is different: elementary strings are split and, therefore, different "mesons" become split too.

In addition to the "mesons" and gauge/quark multiplets, the physical spectrum contains also "baryons" built of N elementary monopoles connected to each other by elementary strings forming a closed "necklace configuration," see Fig. 4.5a. In the classical limit $\tilde{m}^l \gg \Lambda_\sigma$ all strings carry the SU(N) magnetic fluxes given by

$$\int d^2x \, F^*_{\mathrm{SU}(N)} = 2\pi \left(n \cdot n^* - \frac{1}{N} \right), \tag{4.9.3}$$

with $n^l = \delta^{l l_0}$, $l_0 = 1, \dots, N$ for N elementary strings forming the "baryon." The monopoles carry the SU(N) magnetic fluxes given in Eq. (4.5.1) and, therefore, can be located at the corners of the polygon in Fig. 4.5a.

In the highly quantum regime, at $\tilde{m}^l = 0$, both strings and monopoles carry no average SU(N) magnetic flux, see (4.9.1). The confined monopoles are seen as kinks interpolating between the "neighboring" quantum vacua of the CP($N - 1$) model (a.k.a. strings) in the closed necklace configuration in Fig. 4.5a.

As was mentioned, the monopoles/kinks acquire flavor global quantum numbers. They become fundamentals in SU(N)$_{C+F}$. Thus, the "baryon" is in the

$$\prod_1^N (N)$$

representation of SU(N)$_{C+F}$. Note that both quarks and monopoles do not carry baryon numbers. Therefore, our "baryon" has no baryon number too. The reason for this is that the U(1) baryon current is coupled to a gauge boson in the U(N) gauge theory that we consider here. This means, in particular, that the "baryons" can decay into the monopole "mesons" or gauge/quark multiplets.

We mentioned that the "neighboring" elementary strings can form a non-BPS bound state, a composite string. It is plausible then that in practice the monopole "baryon" actually resembles a configuration shown in Fig. 4.5b.

Let us emphasize that all states seen in the physical spectrum of the theory are gauge singlets. This goes without saying. While color charges of the gauge/quark multiplets are screened by the Higgs mechanism, the monopoles are confined by non-Abelian strings.

Let us also stress in conclusion that in the limit $\tilde{m}^l = 0$ the global group SU(N)$_{C+F}$ is restored in the bulk and both strings and confined monopoles become non-Abelian. One might argue that this restoration could happen only at the classical level. One could suspect that in quantum theory a "dynamical Abelization" (i.e. a cascade breaking of the gauge symmetry U(N)$\rightarrow$U(1)^N $\rightarrow$ discrete subgroup) might occur. This could have happened if the adjoint VEV's that classically vanish at $\tilde{m}^l = 0$ (see (4.1.11)) could develop dynamically in quantum theory.

At $\tilde{m}^l \neq 0$ the global SU(N)$_{C+F}$ group is explicitly broken down to U(1)$^{N-1}$ by the quark masses. At $\tilde{m}^l = 0$ this group is classically restored. If it could be proven to be dynamically broken at $\tilde{m}^l = 0$, this would mean a spontaneous symmetry breaking, with obvious consequences, such as the corresponding Goldstone modes.

We want to explain why this cannot and does not happen in the theory at hand. First of all, if a global symmetry is not spontaneously broken at the tree level then it cannot be broken by quantum effects at weak coupling in "isolated" vacua. Second, if the global group SU(N)$_{C+F}$ were broken spontaneously at $\tilde{m}^l = 0$ this would imply massless Goldstone bosons. However, we know that there are no massless states in the spectrum of the bulk theory, see Section 4.1.

Finally, the breaking of SU(N)$_{C+F}$ in the $\tilde{m}^l = 0$ limit would mean that the twisted masses of the world sheet CP($N - 1$) model would not be given by $\tilde{m}^l$; instead they would be shifted, say,

$$\tilde{m}^l_{(\text{tw})} = \tilde{m}^l + c^l \Lambda_{\text{CP}(N-1)},$$

where c^l are some coefficients. In Section 4.6 it was shown [132, 133] that the BPS spectrum of the CP($N - 1$) model on the string should coincide with the BPS spectrum of the four-dimensional bulk theory on the Coulomb branch. The BPS spectrum of the CP($N - 1$) model is determined by $\tilde{m}^l_{(\text{tw})}$ while the BPS spectrum of the bulk theory on the Coulomb branch is determined by $\tilde{m}^l$. In [30] it was shown that the BPS spectra of both theories coincide at $\tilde{m}^l_{(tw)} = \tilde{m}^l$. Thus, we conclude that $c^l = 0$, and the twisted masses vanish in the $\tilde{m}^l = 0$ limit.

Hence, the global $SU(N)_{C+F}$ group is not broken in the bulk and both strings and confined monopoles become non-Abelian at $\tilde{m}^l = 0$.

5

Less supersymmetry

Let us move towards less supersymmetric theories. In this chapter we will review non-Abelian strings in four-dimensional gauge theories with $\mathcal{N} = 1$. In Chapter 6 we will deal with $\mathcal{N} = 0$.

As was discussed in the Introduction to Part II, the Seiberg–Witten mechanism of confinement [2, 3] relies on a cascade gauge symmetry breaking: the non-Abelian gauge group breaks down to an Abelian subgroup at a higher scale by condensation of the adjoint scalars, and at a lower scale the Abelian subgroup breaks down to a discrete subgroup by condensation of quarks (or monopoles, depending on the type of vacuum considered). This leads to formation of the ANO flux tubes and ensures an Abelian nature of confinement of the monopoles (or quarks, respectively). The gauge group acting in the infrared, where the confinement mechanism becomes operative, is Abelian.

On the other hand, non-supersymmetric QCD-like theories as well as $\mathcal{N} = 1$ SQCD have no adjoint scalars and, as a result, no cascade gauge symmetry breaking occurs. The gauge group acting in the infrared is non-Abelian. Confinement in these theories is non-Abelian. This poses a problem of understanding confinement in theories of this type. Apparently, a straightforward extrapolation of the Seiberg–Witten confinement scenario to these theories does not work.

The discovery of the non-Abelian strings [130, 131, 132, 133] suggests a novel possibility of solving this problem. In the $\mathcal{N} = 2$ gauge theory (4.1.7) the SU(N) subgroup of the U(N) gauge group remains unbroken after the squark condensation; the vacuum expectation value $\langle a^a \rangle = 0$, see (4.1.11). This circumstance demonstrates that the formation of the non-Abelian strings does not rely on the presence of adjoint VEVs. This suggests, in turn, that we can give masses to the adjoint fields, make them heavy, and eventually decouple the adjoint fields altogether, without losing qualitative features of the non-Abelian confinement mechanism reviewed above.

This program – moving towards less supersymmetry – was initiated in Ref. [189] which we will discuss in this section. In [189] we considered $\mathcal{N} = 2$ gauge theory (4.1.7), with the gauge group U(N). This theory was deformed by a mass term μ for the adjoint matter fields breaking $\mathcal{N} = 2$ supersymmetry down to $\mathcal{N} = 1$. The breaking terms do not affect classical solutions for the non-Abelian strings. The latter are still 1/2 BPS-saturated. However, at the quantum level the strings "feel" the presence of $\mathcal{N} = 2$ supersymmetry breaking terms. Effects generated by these terms first show up in the sector of the fermion zero modes.

Upon breaking $\mathcal{N} = 2$ supersymmetry in the bulk theory down to $\mathcal{N} = 1$, the fermionic sector of the world sheet theory modifies. The number of the fermion zero modes on the string (and hence the number of the fermion fields in the world sheet theory) does not change. It is determined by the index theorem [104] which we discuss in Section 5.4. However, supersymmetry of the world sheet model changes. The $\mathcal{N} = (2, 2)$ supersymmetry of the undeformed CP($N - 1$) model is broken down to $\mathcal{N} = (0, 2)$ by the bulk mass term μ [190, 191]. Moreover, the superorientational sector of the model gets mixed with the supertranslational one. The world sheet model emerging after deformation is called the heterotic CP($N - 1$) model. We will review in this section how the heterotic world sheet model emerges, as well as the physics of the heterotic $CP(N - 1)$ model which happens to be solvable in the large-N expansion [192]. The solution exhibits supersymmetry breaking at the quantum level.

If the adjoint mass parameter μ is kept finite, the non-Abelian string in the $\mathcal{N} = 1$ model at hand is well-defined and supports confined monopoles. However, at $\mu \to \infty$, as the adjoint superfield becomes very heavy (i.e. we approach the limit of $\mathcal{N} = 1$ SQCD) an infrared problem develops. This is due to the fact that in $\mathcal{N} = 1$ SQCD defined in a standard way the vacuum manifold is no longer an isolated point. Rather, a flat direction develops (a Higgs branch). The presence of the massless states obscures physics of the non-Abelian strings. In particular, the strings become infinitely thick [189]. Thus, one arrives at a dilemma: either one must abandon the attempt to decouple the adjoint super-field, or, if this decoupling is performed, confining non-Abelian strings cease to exist [189].

A way out was suggested in [104]. A relatively insignificant modification of the benchmark $\mathcal{N} = 2$ model cures the infrared problem. All we have to do is to add a neutral meson superfield M coupled to the quark superfields through a super-potential term. Acting together with the mass term of the adjoint superfield, M breaks $\mathcal{N} = 2$ down to $\mathcal{N} = 1$. The limit $\mu \to \infty$ in which the adjoint superfield completely decouples, becomes well-defined. No flat directions emerge. The limiting theory is $\mathcal{N} = 1$ SQCD supplemented by the meson superfield. It supports non-Abelian strings. The junctions of these strings present confined monopoles, or, better to say, what becomes of the monopoles in the theory where there are no adjoint scalar fields. There is a continuous path following which one can trace the monopole evolution in its entirety: from the 't Hooft–Polyakov monopoles which do not exist without the adjoint scalars to the confined monopoles in the adjoint-free environment.

5.1 Breaking $\mathcal{N} = 2$ supersymmetry down to $\mathcal{N} = 1$

In Section 5.1 we will outline main results of Ref. [189, 190, 191] where non-Abelian strings were considered in an $\mathcal{N} = 1$ gauge theory obtained as a deformation of the $\mathcal{N} = 2$ theory (4.1.7) by mass terms of the adjoint matter.

5.1.1 Deformed theory and string solutions

Let us add a superpotential mass term to our $\mathcal{N} = 2$ SQCD,

$$
\mathcal{W}_{\mathcal{N}=1} = \sqrt{\frac{N}{2}} \frac{\mu_1}{2} \mathcal{A}^2 + \frac{\mu_2}{2} (\mathcal{A}^a)^2 , \tag{5.1.1}
$$

where μ_1 and μ_2 are mass parameters for the chiral superfields belonging to $\mathcal{N} = 2$ gauge supermultiplets, U(1) and SU(N), respectively, while the factor $\sqrt{N/2}$ is included for convenience. Clearly, the mass term (5.1.1) splits these supermultiplets, breaking $\mathcal{N} = 2$ supersymmetry down to $\mathcal{N} = 1$.

The bosonic part of the SU(N)$\times$U(1) theory has the form (4.1.7) with the potential

$$
\begin{aligned}
V(q^A, \tilde{q}_A, a^a, a)_{\mathcal{N}=1} = {} & \frac{g_2^2}{2} \left(\frac{1}{g_2^2} f^{abc} \bar{a}^b a^c + \bar{q}_A T^a q^A - \tilde{q}_A T^a \bar{\tilde{q}}^A \right)^2 \\
& + \frac{g_1^2}{8} \left(\bar{q}_A q^A - \tilde{q}_A \bar{\tilde{q}}^A - N\xi \right)^2 \\
& + \frac{g_2^2}{2} \left| 2\tilde{q}_A T^a q^A + \sqrt{2} \mu_2 a^a \right|^2 + \frac{g_1^2}{2} \left| \tilde{q}_A q^A + \sqrt{N} \mu_1 a \right|^2 \\
& + \frac{1}{2} \sum_{A=1}^{N} \left\{ \left| (a + 2T^a a^a) q^A \right|^2 \right. \\
& \left. \qquad\qquad + \left| (a + 2T^a a^a) \bar{\tilde{q}}_A \right|^2 \right\} ,
\end{aligned} \tag{5.1.2}
$$

where the sum over repeated flavor indices A is implied. The potential (5.1.2) differs from the one in (4.1.9) in two ways. First, we use SU(2)$_R$ invariance of the original

$\mathcal{N} = 2$ theory with the potential (4.1.9) to rotate the FI term. In Eq. (5.1.2) it is the FI D term, while in Chapter 4 we considered the FI F term.

Second, there are $\mathcal{N} = 2$ supersymmetry breaking contributions from F terms in Eq. (5.1.2) proportional to the mass parameters μ_1 and μ_2. Note that we set the quark mass differences at zero and redefine a to absorb the average value of the quark mass parameters.

As in Eq. (4.1.9), the FI term triggers spontaneous breaking of the gauge symmetry. The vacuum expectation values of the squark fields can be chosen as

$$\langle q^{kA}\rangle = \sqrt{\xi}\begin{pmatrix} 1 & 0 & \ldots \\ \ldots & \ldots & \ldots \\ \ldots & 0 & 1 \end{pmatrix}, \quad \langle \bar{\tilde{q}}^{kA}\rangle = 0,$$

$$k = 1,\ldots,N, \quad A = 1,\ldots,N, \tag{5.1.3}$$

while the adjoint field VEVs are

$$\langle a^a \rangle = 0, \quad \langle a \rangle = 0, \tag{5.1.4}$$

see (4.1.11).

We see that the quark VEVs have the color-flavor locked form (see (4.1.14)) implying that the SU$(N)_{C+F}$ global symmetry is unbroken in the vacuum. Much in the same way as in $\mathcal{N} = 2$ SQCD, this symmetry leads to the emergence of the orientational zero modes of the Z_N strings.

Note that VEVs (5.1.3) and (5.1.4) do not depend on supersymmetry breaking parameters μ_1 and μ_2. This is due to the fact that our choice of parameters in (5.1.2) ensures vanishing of the adjoint VEVs, see (5.1.4). In particular, we have the same pattern of symmetry breaking all the way up to very large μ_1 and μ_2, where the adjoint fields decouple. As in $\mathcal{N} = 2$ SQCD we assume $\sqrt{\xi} \gg \Lambda_{SU(2)}$ to ensure weak coupling.

Now, let us discuss the mass spectrum in the $\mathcal{N} = 1$ theory at hand. Since both U(1) and SU(N) gauge groups are broken by squark condensation, all gauge bosons become massive. Their masses are given in Eqs. (4.1.16) and (4.1.17).

To obtain the scalar boson masses we expand the potential (5.1.2) near the vacuum (5.1.3), (5.1.4) and diagonalize the corresponding mass matrix. Then, N^2 components of $2N^2$ (real) component scalar field q^{kA} are eaten by the Higgs mechanism for the U(1) and SU(N) gauge groups. Other N^2 components are split as follows. One component acquires mass (4.1.17) and becomes the scalar component of a massive $\mathcal{N} = 1$ vector U(1) gauge multiplet. Moreover, $N^2 - 1$ components acquire masses (4.1.16) and become scalar superpartners of the SU(N) gauge bosons in $\mathcal{N} = 1$ massive gauge supermultiplets.

Other $4N^2$ real scalar components of the fields $\tilde{q}_{Ak}$, a^a and a produce the following states: two states acquire mass

$$m^+_{U(1)} = g_1 \sqrt{\frac{N}{2} \xi}\, \lambda^+_1 \,, \tag{5.1.5}$$

while the mass of other two states is given by

$$m^-_{U(1)} = g_1 \sqrt{\frac{N}{2} \xi}\, \lambda^-_1 \,, \tag{5.1.6}$$

where $\lambda^\pm_1$ are two roots of the quadratic equation

$$\lambda^2_i - \lambda_i (2 + \omega^2_i) + 1 = 0 \,, \tag{5.1.7}$$

for $i = 1$, where we introduced two $\mathcal{N} = 2$ supersymmetry breaking parameters $\omega_{1,2}$ associated with the U(1) and SU(N) gauge groups, respectively,

$$\omega_1 = \frac{g_1 \mu_1}{\sqrt{\xi}} \,, \quad \omega_2 = \frac{g_2 \mu_2}{\sqrt{\xi}} \,. \tag{5.1.8}$$

Other $2(N^2 - 1)$ states acquire mass

$$m^+_{SU(N)} = g_2 \sqrt{\xi}\, \lambda^+_2 \,, \tag{5.1.9}$$

while the remaining $2(N^2 - 1)$ states become massive, with mass

$$m^-_{SU(N)} = g_2 \sqrt{\xi}\, \lambda^-_2 \,, \tag{5.1.10}$$

where $\lambda^\pm_2$ are two roots of the quadratic equation (5.1.7) for $i = 2$. Note that all states come either as singlets or adjoints with respect to the unbroken SU(N)$_{C+F}$.

When the SUSY breaking parameters ω_i vanish, the masses (5.1.5) and (5.1.6) coincide with the U(1) gauge boson mass (4.1.17). The corresponding states form the bosonic part of a long $\mathcal{N} = 2$ massive U(1) vector supermultiplet [35], see also Section 4.1.2.

If $\omega_1 \neq 0$ this supermultiplet splits into an $\mathcal{N} = 1$ vector multiplet, with mass (4.1.17), and two chiral multiplets, with masses (5.1.5) and (5.1.6). The same happens with the states with masses (5.1.9) and (5.1.10). In the limit of vanishing ω's they combine into bosonic parts of $(N^2 - 1)$ $\mathcal{N} = 2$ vector supermultiplets with mass (4.1.16). If $\omega_i \neq 0$ these multiplets split into $(N^2 - 1)$ $\mathcal{N} = 1$ vector multiplets (for the SU(N) group) with mass (4.1.16) and $2(N^2 - 1)$ chiral multiplets with masses (5.1.9) and (5.1.10). Note that the same splitting pattern was found in [35] in the Abelian case.

Now let us take a closer look at the spectrum obtained above in the limit of large $\mathcal{N} = 2$ supersymmetry breaking parameters ω_i, $\omega_i \gg 1$. In this limit the larger masses $m^+_{U(1)}$ and $m^+_{SU(N)}$ become

$$m^+_{U(1)} = m_{U(1)}\omega_1 = g_1^2\sqrt{\frac{N}{2}}\mu_1 , \quad m^+_{SU(N)} = m_{SU(N)}\omega_2 = g_2^2\mu_2 . \quad (5.1.11)$$

In the limit $\mu_i \to \infty$ these are the masses of the heavy adjoint scalars a and a^a. At $\omega_i \gg 1$ these fields decouple and can be integrated out.

The low-energy theory in this limit contains massive gauge $\mathcal{N} = 1$ multiplets and chiral multiplets with the lower masses m^-. Equation (5.1.7) gives for these masses

$$m^-_{U(1)} = \frac{m_{U(1)}}{\omega_1} = \sqrt{\frac{N}{2}}\frac{\xi}{\mu_1} , \quad m^-_{SU(2)} = \frac{m_{SU(2)}}{\omega_2} = \frac{\xi}{\mu_2} . \quad (5.1.12)$$

In particular, in the limit of infinite μ_i these masses tend to zero. This reflects the presence of a Higgs branch in $\mathcal{N} = 1$ SQCD. To see the Higgs branch and calculate its dimension, please observe that our theory (4.1.7) with the potential (5.1.2) in the limit $\mu_i \to \infty$ flows to $\mathcal{N} = 1$ SQCD with the gauge group $SU(N)\times U(1)$ and the FI D term. The bosonic part of the action of the latter theory is

$$
S = \int d^4x \left\{ \frac{1}{4g_2^2}(F^a_{\mu\nu})^2 + \frac{1}{4g_1^2}(F_{\mu\nu})^2 + |\nabla_\mu q^A|^2 + |\nabla_\mu \tilde{\bar{q}}^A|^2 \right.
$$
$$
\left. + \frac{g_2^2}{2}(\bar{q}_A T^a q^A - \tilde{q}_A T^a \tilde{\bar{q}}^A)^2 + \frac{g_1^2}{8}(\bar{q}_A q^A - \tilde{q}_A \tilde{\bar{q}}^A - N\xi)^2 \right\} \quad (5.1.13)
$$

All F terms disappear in this limit, and we are left with the D terms. We have $4N^2$ real components of the q and $\tilde{q}$ fields while the number of the D term constraints in (5.1.13) is N^2. Moreover, N^2 phases are eaten by the Higgs mechanism. Thus, the dimension of the Higgs branch in Eq. (5.1.13) is

$$4N^2 - N^2 - N^2 = 2N^2 .$$

The vacuum (5.1.3) corresponds to the base point of the Higgs branch with $\tilde{q} = 0$. In other words, flowing from $\mathcal{N} = 2$ theory (4.1.7), we do not recover the entire Higgs branch of $\mathcal{N} = 1$ SQCD. Instead, we arrive at a single vacuum – a base point of the Higgs branch.

The scale of $\mathcal{N} = 1$ SQCD

$$\Lambda^{\mathcal{N}=1}_{SU(N)}$$

is expressed in terms of the scale $\Lambda_{SU(N)}$ of the deformed $\mathcal{N} = 2$ theory as follows:

$$\left(\Lambda_{SU(N)}^{\mathcal{N}=1}\right)^{2N} = \mu_2^N \, \Lambda_{SU(N)}^N \,. \tag{5.1.14}$$

To keep the bulk theory at weak coupling in the limit of large μ_i we assume that

$$\sqrt{\xi} \gg \Lambda_{SU(N)}^{\mathcal{N}=1} \,. \tag{5.1.15}$$

Now, considering the theory (4.1.7) with the potential (5.1.2), let us return to the case of arbitrary μ_i and discuss non-Abelian string solutions. The BPS saturation is maintained. By the same token, as for the BPS strings in $\mathcal{N} = 2$ we use the *ansatz*

$$q^{kA} \equiv \varphi^{kA}, \quad \tilde{q}_{Ak} = 0\,. \tag{5.1.16}$$

The adjoint fields are set to zero. Note that Eq. (5.1.16) is an $SU(2)_R$-rotated version of (4.2.1). The FI F term considered in Chapter 4 is rotated into the FI D term in (5.1.2).

With these simplifications the $\mathcal{N} = 1$ model (4.1.7) with the potential (5.1.2) reduces to the model (4.2.2) which was exploited in Chapter 4 to obtain non-Abelian string solutions. The reason for this is that the adjoint fields play no role in the string solutions, and we let them vanish, see Eq. (5.1.4). Then $\mathcal{N} = 2$ breaking terms vanish, and the potential (5.1.2) reduces to the one in Eq. (4.1.9) (up to an $SU(2)_R$ rotation).

This allows us to parallel the construction of the non-Abelian strings carried out in Section 4.3. In particular, the elementary string solution is given by Eq. (4.4.4). Moreover, the *bosonic* part of the world sheet theory is nothing but the $CP(N - 1)$ sigma model (4.4.9), with the coupling constant β determined by the coupling g_2 of the bulk theory via Eq. (4.4.15) at the scale $\sqrt{\xi}$. The latter scale plays the role of the UV cut off in the world sheet theory.

At small values of the deformation parameter,

$$\mu_2 \ll \sqrt{\xi}\,,$$

the coupling constant g_2 of the four-dimensional bulk theory is determined by the scale $\Lambda_{SU(N)}$ of the $\mathcal{N} = 2$ theory. Then Eq. (4.4.15) implies (see (4.4.35))

$$\Lambda_\sigma = \Lambda_{SU(N)}\,, \tag{5.1.17}$$

where we take into account that the first coefficient of the β function is N both in the $\mathcal{N} = 2$ limit of the four-dimensional bulk theory and in the two-dimensional $CP(N - 1)$ model, see (4.4.34).

Instead, in the limit of large μ_2,

$$\mu_2 \gg \sqrt{\xi},$$

the coupling constant g_2 of the bulk theory is determined by the scale $\Lambda_{SU(N)}^{\mathcal{N}=1}$ of $\mathcal{N} = 1$ SQCD (5.1.13), see (5.1.14). In this limit Eq. (4.4.15) gives

$$\Lambda_\sigma = \frac{\left(\Lambda_{SU(N)}^{\mathcal{N}=1}\right)^2}{g_2 \sqrt{\xi}}, \tag{5.1.18}$$

where we take into account the fact that the first coefficient of the β function in $\mathcal{N} = 1$ SQCD is $2N$.

5.1.2 Heterotic CP(N − 1) model

In this section we will discuss the fermionic sector of the low-energy effective theory on the world sheet of the non-Abelian string in the deformed bulk theory (4.1.7) with the potential (5.1.2), as well as supersymmetry of the world sheet theory. First, we note that our string is classically 1/2 BPS-saturated. Therefore, in the $\mathcal{N} = 2$ limit (with $\mathcal{N} = 2$ breaking parameters μ_i vanishing) four supercharges out of eight present in the bulk theory are automatically preserved on the string world sheet. They become supercharges in the CP(N − 1) model.

What happens when we break $\mathcal{N} = 2$ supersymmetry of the bulk model by switching on parameters μ_i (for simplicity we consider the case $\mu_1 = \mu_2 \equiv \mu$ here)? The 1/2 "BPS-ness" of the string solution requires only two supercharges on the world sheet. However, as we will show in Section 5.4, the number of the fermion zero modes in the string background does not change. This number is fixed by the index theorem. Thus, the number of (classically) massless fermion fields in the world sheet model does not change.

It is well known that the $\mathcal{N} = (2, 2)$ supersymmetric sigma model with the CP(N − 1) target space does not admit $\mathcal{N} = (0, 2)$ supersymmetric deformations [156]. A way out was indicated in [190]: the world-sheet theory is in fact CP(N − 1) × C rather than the CP(N − 1) model. The factor C comes from the translational sector. In the $\mathcal{N} = 2$ limit the translational and the orientational sectors of the world-sheet theory are totally decoupled. Breaking $\mathcal{N} = 2$ supersymmetry in the bulk mixes fermions from these two sectors on the world sheet.

The translational sector of the effective theory on the string in the $\mathcal{N} = 2$ limit contains the bosonic field x_{0i}, $i = 1, 2$ (position of string's center in the (1,2) plane), and two fermion fields ζ_L and ζ_R. Two supercharges that survive on the string world sheet at non-zero μ protect x_{0i} and ζ_L. The world-sheet fields $x_{0i}(t, z)$ and $\zeta_L(t, z)$

remain free fields decoupled from all others. This is no longer the case with regards to ζ_R which gets an interaction with fermions of the orientational sector.

As a result, the heterotic $\mathcal{N} = (0, 2)$ model in the gauged formulation (see (4.4.32)) takes the form [190]

$$
\begin{aligned}
S = \int d^2x \Bigg\{ & \frac{1}{2} \bar{\zeta}_R \, i\partial_L \, \zeta_R + \left[2\sqrt{\beta} \, i \, \delta \, \bar{\lambda}_L \, \zeta_R + \text{H.c.} \right] \\
& + |\nabla_k n^l|^2 + \frac{1}{4e^2} F_{kl}^2 + \frac{1}{e^2} |\partial_k \sigma|^2 + \frac{1}{2e^2} D^2 + 2|\sigma|^2 |n^l|^2 + i D(|n^l|^2 - 2\beta) \\
& + \bar{\xi}_{lR} \, i\nabla_L \, \xi_R^l + \bar{\xi}_{lL} \, i\nabla_R \, \xi_L^l + \frac{1}{e^2} \bar{\lambda}_R \, i\partial_L \, \lambda_R + \frac{1}{e^2} \bar{\lambda}_L \, i\partial_R \, \lambda_L \\
& + \left[i\sqrt{2} \sigma \, \bar{\xi}_{lR} \xi_L^l + i\sqrt{2} \bar{n}_l (\lambda_R \xi_L^l - \lambda_L \xi_R^l) + \text{H.c.} \right] + 8\beta \, |\delta|^2 \, |\sigma|^2 \Bigg\} , \quad (5.1.19)
\end{aligned}
$$

where we omitted the fields $x_{0i}(t, z)$ and $\zeta_L(t, z)$ as irrelevant for the present consideration, while $\partial_{L,R} = \partial_0 \mp i \, \partial_3$. Here $\xi_{R,L}^l$ are fermionic superpartners of the bosonic orientational fields n^l. For convenience we change the normalization of n^l as compared with the one in (4.4.32) absorbing the coupling constant 2β in the kinetic term for n's.

Much in the same way as the $\mathcal{N} = (2, 2)$ CP($N - 1$) model, this heterotic model should be considered in the strong coupling limit $e^2 \to \infty$. The gauge multiplet consists of the U(1) gauge field A_k, complex scalar σ, fermions $\lambda_{R,L}$ and axillary field D. Integrating over D gives constraint (4.4.3) modified due to the change of normalization of n as follows:

$$
|n^l|^2 = 2\beta . \quad (5.1.20)
$$

The terms in Eq. (5.1.19) containing the deformation parameter δ break $\mathcal{N} = (2, 2)$ supersymmetry down to $\mathcal{N} = (0, 2)$. The parameter δ is complex and dimensionless. It was calculated in terms of the deformation parameter μ of the bulk theory in [191]. We review this result below.

Integrating over the axillary fields λ we arrive at the constraints

$$
\bar{n}_l \, \xi_L^l = 0, \quad \bar{\xi}_{lR} \, n^l = \sqrt{2\beta} \, \delta \, \zeta_R , \quad (5.1.21)
$$

replacing those in Eq. (4.4.24) (the latter equation in the gauged formulation for arbitrary N takes the form $\bar{n}_l \, \xi^l = 0$). We see that the constraint (4.4.24) is modified for the right-handed fermions ξ_R implying that the supertranslational sector of the world sheet theory is no longer decoupled from the orientational one. The general structure of the deformation in (5.1.19) is dictated by $\mathcal{N} = (0, 2)$ supersymmetry.

Integrating over A_k and σ produces four-fermion interactions in the model (5.1.19). Once the coefficient in front of $|\sigma|^2$ is modified by the $\mathcal{N} = 2$ breaking deformation the coefficient in front of the four-fermion interaction is modified as well (as compared with Eq. (4.4.31)).

The model (5.1.19) has a U(1) axial symmetry which is broken by the chiral anomaly down to the discrete subgroup Z_{2N} [159]. The σ field is related to the fermion bilinear operator by the following formula:

$$\sigma = -\frac{i}{2\sqrt{2}\beta(1 + 2|\delta|^2)} \, \bar{\xi}_{IL}\xi_R^l \, . \tag{5.1.22}$$

Moreover, it transforms under the above Z_{2N} symmetry as

$$\sigma \to e^{\frac{2\pi k}{N} i} \sigma, \quad k = 1, \ldots, N - 1 \, . \tag{5.1.23}$$

We will see below that the Z_{2N} symmetry is spontaneously broken by a condensate of σ, down to Z_2, much in the same way as in the conventional $\mathcal{N} = (2, 2)$ model [159]. This is equivalent to saying that the fermion bilinear condensate $\langle \bar{\xi}_{IL}\xi_R^l \rangle$ develops, breaking the discrete Z_{2N} symmetry down to Z_2.

We can rewrite the indirect interactions between superorientational and super-translational sectors of the theory coded in the constraint (5.1.21) by shifting the field ξ_R as follows:

$$\bar{\xi}_R \to \bar{\xi}_R - \delta \frac{1}{\sqrt{2\beta}} \bar{n} \, \zeta_R \, . \tag{5.1.24}$$

Then we return to unmodified constraints

$$\bar{n}_l \, \xi_L^l = 0, \qquad \bar{\xi}_{lR} \, n^l = 0 \, . \tag{5.1.25}$$

Performing a rather straightforward algebraic analysis based on the relation between the gauged and O(3) formulations (see Appendix B.4) we get

$$
\begin{aligned}
S_{1+1} = \beta \int d^2x \, \Big\{ &\frac{1}{2}(\partial_k S^a)^2 + \frac{1}{2} \chi_R^a \, i \, \partial_L \, \chi_R^a + \frac{1}{2} \chi_L^a \, i \, \partial_R \, \chi_L^a - \frac{c^2}{2}(\chi_R^a \chi_L^a)^2 \\
&+ \frac{c}{2\sqrt{2}\beta} \chi_R^a \big(i \, \partial_L \, S^a \, (\alpha \, \zeta_R + \bar{\alpha} \, \bar{\zeta}_R) + i\varepsilon^{abc} S^b i \, \partial_L \, S^c \, (\alpha \, \zeta_R - \bar{\alpha} \, \bar{\zeta}_R)\big) \\
&+ \frac{1}{2\beta} \bar{\zeta}_R \, i \, \partial_L \, \zeta_R + |\alpha|^2 \frac{c^2}{4\beta} \bar{\zeta}_R \zeta_R \, i\varepsilon^{abc} S^a \chi_L^b \chi_L^c \Big\} \, ,
\end{aligned}
\tag{5.1.26}
$$

where

$$c^2 = \frac{1}{1 + |\alpha|^2}, \tag{5.1.27}$$

we restrict ourselves to $N = 2$ and rename $\chi_{1,2}^a$ of Section 4.4.2, $\chi_{1,2}^a \to \chi_{R,L}^a$.

The relation of the deformation parameter introduced here and the one in the gauged formulation of the theory (see Eqs. (5.1.19)) is as follows:

$$\delta = \frac{\alpha}{\sqrt{1 - |\alpha|^2}}. \tag{5.1.28}$$

As a result of the shift (5.1.24), crucial bifermionic terms of the type $\chi_R \partial_L S \zeta_R$ appear in the second line of Eq. (5.1.26).

Now the problem is to actually derive the heterotic world sheet CP($N - 1$) model from the bulk theory. The general structure of the theory (5.1.19) is fixed by $\mathcal{N} = (0, 2)$ supersymmetry. In order to derive this theory as an effective low-energy theory on the string we have to calculate the deformation parameter δ in terms of the bulk parameters. The kinetic cross-terms $\chi_R \partial S \zeta_R$ in the formulation (5.1.26), bilinear in the fermion fields, allow us to do so. In Ref. [191] the μ-deformation of the fermion zero modes was considered. Both supertranslational and superorientational fermion zero modes on the string were found in the limits of small and large μ by solving the Dirac equations. The overlap of the translational and orientational fermion zero modes gives the kinetic cross-term $\chi_1 \partial S \zeta_R$ in (5.1.19).

This derivation provides us with a relation between the bulk and world sheet deformation parameters, namely [191],

$$\delta = \begin{cases} \text{const } \dfrac{g_2^2 \mu}{M_{\mathrm{SU}(N)}}, & \text{small } \mu, \\[2em] \text{const } \dfrac{\mu}{|\mu|} \sqrt{\ln \dfrac{g_2^2 |\mu|}{M_{\mathrm{SU}(N)}}}, & \text{large } \mu, \end{cases} \tag{5.1.29}$$

where the mass of the gauge boson $M_{\mathrm{SU}(N)}$ is given in Eq. (4.1.16). The constants here are determined by the profile functions of the string solution [191].

The physical reason for the logarithmic behavior of the world sheet deformation parameter at large μ is as follows. In the large-μ limit certain states in the bulk theory become light [189, 191], see Section 5.1.1. This reflects the presence of the Higgs branch in $\mathcal{N} = 1$ SQCD to which our bulk theory flows in the $\mu \to \infty$ limit. The argument of the logarithm in (5.1.29) is the ratio of $M_{\mathrm{SU}(N)}$ and the small mass of the light states associated with this would-be Higgs branch [191].

To conclude this section, let us mention that more general deformations of $\mathcal{N} = 2$ theory (4.1.7) preserving $\mathcal{N} = 1$ supersymmetry were also considered in [190, 191]. In particular, deformations of (4.1.7) with unequal quark masses with a polynomial superpotential

$$\mathcal{W} = \mathrm{Tr} \sum_{k=1}^{N} \frac{c_k}{k+1} \Phi^{k+1} \tag{5.1.30}$$

do not spoil the BPS nature of string solutions if the critical points of the super-potential coincide with the quark mass parameters. Associated $\mathcal{N} = (0, 2)$ heterotic deformations of the CP($N - 1$) world-sheet theory were derived in [191]. For polynomial deformations (5.1.30) a non-polynomial (logarithmic) response was found in the world-sheet model. The heterotic CP($N - 1$) model in the geometric formulation is presented in Appendix B.5.

5.1.3 Large-N solution

The $\mathcal{N} = (2, 2)$ model as well as the non-supersymmetric $CP(N - 1)$ model were solved by Witten in the large-N limit [159]. The same method was used in [192] to study the $\mathcal{N} = (0, 2)$ heterotic $CP(N - 1)$ model. In this section we will briefly review this analysis.

Since the action (5.1.19) is quadratic in the fields n^l and ξ^l we can integrate out these fields and then minimize the resulting effective action with respect to the fields from the gauge multiplet. The large-N limit ensures the corrections to the saddle point approximation to be small. In fact, this procedure boils down to calculating a small set of one-loop graphs with the n^l and ξ^l fields propagating in loops. After integrating n^l and ξ^l out, we must check self-consistency.

Integration over n^l and ξ^l in (5.1.19) yields the following determinants:

$$\left[\det\left(-\partial_k^2 + iD + 2|\sigma|^2\right)\right]^{-N}\left[\det\left(-\partial_k^2 + 2|\sigma|^2\right)\right]^{N}, \qquad (5.1.31)$$

where we dropped the gauge field A_k. The first determinant here comes from the boson loops while the second from fermion loops. Note, that the n^l mass is given by $iD + 2|\sigma|^2$ while that of the fermions ξ^l is $2|\sigma|^2$. If supersymmetry is unbroken (i.e. $D = 0$) these masses are equal, and the product of the determinants reduces to unity, as it should be.

Calculation of the determinants in Eq. (5.1.31) is straightforward. We easily get the following contribution to the effective action:

$$\frac{N}{4\pi}\left\{\left(iD + 2|\sigma|^2\right)\left[\ln\frac{M_{\mathrm{uv}}^2}{iD + 2|\sigma|^2} + 1\right] - 2|\sigma|^2\left[\ln\frac{M_{\mathrm{uv}}^2}{2|\sigma|^2} + 1\right]\right\}, \qquad (5.1.32)$$

where quadratically divergent contributions from bosons and fermions do not depend on D and σ and cancel each other. Here M_{uv} is an ultraviolet cut off. Remembering that the action in (5.1.19) presents an effective low-energy theory on the string world sheet one can readily identify the UV cut off in terms of the bulk parameters,

$$M_{\mathrm{uv}} = M_{\mathrm{SU}(N)}. \qquad (5.1.33)$$

Invoking Eq. (4.4.15) we conclude that the bare coupling constant β in (5.1.19) can be parameterized as

$$\beta = \frac{N}{8\pi} \ln \frac{M_{uv}^2}{\Lambda_\sigma^2}. \tag{5.1.34}$$

Substituting this expression in (5.1.19) and adding the one-loop correction (5.1.32) we see that the term proportional to $iD \ln M_{uv}^2$ is canceled out, and the effective action is expressed in terms of the renormalized coupling constant,

$$\beta_{\text{ren}} = \frac{N}{8\pi} \ln \frac{iD + 2|\sigma|^2}{\Lambda_\sigma^2}. \tag{5.1.35}$$

Assembling all contributions together we get the effective potential as a function of the D and σ fields in the form

$$V_{\text{eff}} = \int d^2x \, \frac{N}{4\pi} \left\{ -\left(iD + 2|\sigma|^2\right) \ln \frac{iD + 2|\sigma|^2}{\Lambda_\sigma^2} + iD \right.$$
$$\left. + 2|\sigma|^2 \ln \frac{2|\sigma|^2}{\Lambda_\sigma^2} + 2|\sigma|^2 u \right\}, \tag{5.1.36}$$

where instead of the deformation parameter δ we introduced a more convenient (dimensionless) parameter u which does not scale with N,

$$u = \frac{16\pi}{N} \beta \, |\delta|^2. \tag{5.1.37}$$

Minimizing this potential with respect to D and σ we arrive at the following relations:

$$\beta_{\text{ren}} = \frac{N}{8\pi} \ln \frac{iD + 2|\sigma|^2}{\Lambda_\sigma^2} = 0,$$
$$\ln \frac{iD + 2|\sigma|^2}{2|\sigma|^2} = u. \tag{5.1.38}$$

Equations (5.1.38) represent the *master set* which determines the vacua of the theory. Solutions can be readily found,

$$2|\sigma|^2 = \Lambda_\sigma^2 \, e^{-u}, \qquad \sigma = \frac{1}{\sqrt{2}} \Lambda_\sigma \exp\left(-\frac{u}{2} + \frac{2\pi \, i \, k}{N}\right), \qquad k = 0, \ldots, N-1,$$
$$iD = \Lambda_\sigma^2 \left(1 - e^{-u}\right). \tag{5.1.39}$$

The phase factor of σ does not follow from (5.1.38), but we know of its existence from the fact of the spontaneous breaking of the discrete chiral Z_{2N} down to Z_2, see the discussion in Section 5.1.2. Substituting this solution in Eq. (5.1.36) we get the expression for the vacuum energy density [192],

$$\mathcal{E}_{\text{vac}} = \frac{N}{4\pi} i D = \frac{N}{4\pi} \Lambda_\sigma^2 \left(1 - e^{-u}\right). \tag{5.1.40}$$

We see that the vacuum energy does not vanish! The $\mathcal{N} = (0, 2)$ supersymmetry is spontaneously broken. The breaking of $\mathcal{N} = (0, 2)$ supersymmetry was first conjectured in Ref. [193]. Of course, the large-N solution presented above is the most clear-cut demonstration of its spontaneous breaking. However, even in the absence of this solution one can present general arguments in favor of this scenario. Indeed, addition of the extra right-handed field ζ_R in the CP($N-1$) model changes Witten's index from N to zero [191].

Another argument comes from the bulk theory in the limit of large deformation parameter μ. The breaking of supersymmetry for the $\mathcal{N} = (0, 2)$ world-sheet theory could be argued from consistency with the absence of localized BPS solutions of the required type in $\mathcal{N} = 1$ gauge theories. The argument could go along the following lines: If the $\mathcal{N} = (0, 2)$ world-sheet theory is to have supersymmetric vacua, it would need to have N degenerate vacua due to the breaking of the discrete Z_N symmetry. This is likely to imply the existence of BPS kinks preserving one supercharge. From the bulk point of view, these configurations would be 1/4 BPS-saturated confined monopoles. However, such solutions are not supported by the allowed central charges in the $\mathcal{N} = 1$ superalgebra and, therefore, are not expected. Thus, the breaking of world sheet supersymmetry is consistent with the absence of such bulk BPS configurations.

Needless to say, the linear N dependence of the vacuum energy we see in Eq. (5.1.40) was expected.

It is instructive to discuss the first condition in (5.1.38). That $\beta_{\text{ren}} = 0$ was a result of Witten's analysis [159] too. This fact, $\beta_{\text{ren}} = 0$, implies that in quantum theory (unlike the classical one)

$$\langle |n^l|^2 \rangle = 0, \tag{5.1.41}$$

i.e. the global SU(N) symmetry is not spontaneously broken in the vacuum and, hence, there are no massless Goldstone bosons. All bosons get a mass.

If the deformation parameter u vanishes, the vacuum energy vanishes too and supersymmetry is not broken, in full accord with Witten's analysis [159] and with the fact that the Witten index is N in this case [123]. The σ field develops a vacuum expectation value (5.1.39) breaking Z_{2N} symmetry.[1] As we switch on the

---

[1] The vacuum structure (5.1.39) of the $\mathcal{N} = (2, 2)$ model at $u = 0$ was also obtained by Witten for arbitrary N in [157] using a superpotential of the Veneziano–Yankielowicz type [194].

deformation parameter u, the D component develops a VEV; hence, $\mathcal{N} = (0, 2)$ supersymmetry is spontaneously broken. The vacuum energy density no longer vanishes.

In the limit $\mu \to \infty$, the deformation parameter u behaves logarithmically with μ,

$$u = \text{const} \left(\ln \frac{M_{\mathrm{SU}(N)}}{\Lambda_{\mathrm{SU}(N)}^{\mathcal{N}=1}} \right) \left(\ln \frac{g_2^2 |\mu|}{M_{\mathrm{SU}(N)}} \right), \qquad (5.1.42)$$

where the constant above does not depend on N. At any finite u the σ-field condensate does not vanish, labeling N distinct vacua as indicated in Eq. (5.1.39). In each vacuum Z_{2N} symmetry is spontaneously broken down to Z_2. As we explain in Section 5.1.4 we cannot trust the world-sheet theory at very large values of μ due to the presence of the Higgs branch in $\mathcal{N} = 1$ SQCD. Therefore, the vacuum structure outlined above persists in the whole window of the allowed values of the deformation parameter u.

The mass spectrum of the heterotic $\mathrm{CP}(N-1)$ model was determined in Ref. [192] in the large-N limit. In the regime of large u the masses of the n^l bosons and ξ^l fermions become drastically different. They are

$$m_n = \sqrt{iD + 2|\sigma|^2} = \Lambda_\sigma, \quad m_\xi = \sqrt{2}|\sigma| = \Lambda_\sigma \exp\left(-\frac{u}{2}\right), \quad (5.1.43)$$

where we used Eqs. (5.1.39). The fermions are much lighter than their bosonic counterparts.

Much in the same way as in the $\mathcal{N} = (2, 2)$ $\mathrm{CP}(N - 1)$ model [159], the fields belonging to the U(1) gauge multiplet, introduced as axillary fields in (5.1.19), acquire kinetic terms at one loop and become dynamical. Moreover, the photon acquires mass proportional to the VEV of the σ field due to the chiral anomaly. The σ field also becomes massive, with mass determined by $\langle \sigma \rangle$ in (5.1.39).

Due to the spontaneous supersymmetry breaking the theory always has a massless Goldstino. At small u its role is played by ζ_R with a small admixture of other fermions, while in the large u limit the gaugino λ_R becomes massless. In the large u limit when $\langle \sigma \rangle$ is small the low-energy effective theory contains the light (but massive!) photon, two light σ states and only one fermion: the massless Goldstino λ_R.

As was shown above, in the $\mathcal{N} = (0, 2)$ theory supersymmetry is spontaneously broken. The vacuum energy density does not vanish, see (5.1.40). This means that strings under consideration are no longer BPS and their tensions get a shift (5.1.40) with respect to the classical value $T_{\mathrm{cl}} = 2\pi \xi$. However, this shift is the same for all N elementary strings. Their tensions are strictly degenerate; Z_{2N} symmetry is spontaneously broken down to Z_2. The order parameter (the σ field VEV) remains nonvanishing at any finite value of the bulk parameter μ.

The kinks that interpolate between different vacua of the world sheet theory are described by the n^l fields. Their masses are given in Eq. (5.1.43). In the $\mathcal{N} = (0, 2)$ theory the masses of the boson and fermion superpartners are split. The bosonic kinks have masses $\sim \Lambda_\sigma$ in the large-μ limit, while the fermionic kinks become light. Still their masses remain finite and nonvanishing at any finite μ.

We already know that, from the standpoint of the bulk theory, these kinks are confined monopoles [189, 104, 192]. The fact that tensions of all elementary strings are the same ensures that these monopoles are free to move along the string, since with their separation increasing, the energy of the configuration does not change. This means they are in the deconfinement phase on the string.[2] The kinks are deconfined both in $\mathcal{N} = (2, 2)$ and $\mathcal{N} = (0, 2)$ $CP(N-1)$ theories. In other words, individual kinks are present in the physical spectrum of (1+1) dimensional theory. The monopoles, although attached to strings, are free to move on the strings. We will see in Chapter 6 that this is not the case for monopoles in non-supersymmetric theories. Kinks in non-supersymmetric CP($N-1$) models are in the confinement phase on the string, therefore a monopole and an antimonopole attached to the string come close to each other forming a meson-like configuration.

5.1.4 Limits of applicability

As was discussed above, both the string solution and the bosonic part of the world sheet theory for the non-Abelian strings in $\mathcal{N} = 1$ with the potential (5.1.2) are identical to those in $\mathcal{N} = 2$. However, the occurrence of the Higgs branch in the limit $\mu \to \infty$ manifests itself at the quantum level [189]. At the classical level light fields appearing in the bulk theory in the large-μ limit do not enter the string solution. The string is "built" of heavy fields. However, at the quantum level couplings to the light fields lead to a dramatic effect: an effective string thickness becomes large due to long-range tails of the string profile functions associated with the light fields. As a matter of fact, we demonstrated [189, 191] that in the fermion sector this effect is seen already at the classical level. Some of the fermion zero modes on the string solution acquire long-range tails and become non-normalizable in the limit $\mu \to \infty$.

Below we will estimate the range of validity of the description of non-Abelian string dynamics by the CP($N-1$) model (5.1.19). To this end let us note that higher derivative corrections to (5.1.19) run in powers of

$$\Delta\, \partial_k \,, \qquad\qquad\qquad (5.1.44)$$

---

[2] We stress that these monopoles are confined in the bulk theory being attached to strings.

where Δ is the string transverse size (thickness). At small μ it is quite clear that $\Delta \sim 1/g\sqrt{\xi}$. A typical energy scale on the string world sheet is given by the scale Λ_σ of the CP($N-1$) model which, in turn, is given by (5.1.17) at small μ. Thus, $\partial \to \Lambda_\sigma$, and higher-derivative corrections run in powers of $\Lambda_\sigma/g\sqrt{\xi}$. At small μ the higher-derivative corrections are suppressed by powers of $\Lambda_\sigma/g\sqrt{\xi} \ll 1$ and can be ignored. However, with μ increasing, the fermion zero modes acquire long-range tails [189, 191]. This means that an effective thickness of the string grows. The thickness is determined by masses of the lightest states (5.1.12) of the bulk theory,

$$\Delta \sim \frac{1}{m^-} = \frac{\mu}{\xi}.$$ (5.1.45)

The higher-derivative terms are small if $(\Delta \Lambda_\sigma) \ll 1$. Substituting here the scale of the CP($N-1$) model given at large μ by (5.1.18) and the scale of $\mathcal{N}=1$ SQCD (5.1.14) we arrive at the constraint

$$\mu \ll \mu^*,$$ (5.1.46)

where the critical value of μ is

$$g_2^2 \mu^* = \frac{g_2^2 \xi}{\Lambda_\sigma} = \frac{M_{\text{SU}(N)}^3}{\left(\Lambda_{\text{SU}(N)}^{\mathcal{N}=1}\right)^2}.$$ (5.1.47)

If the condition (5.1.46) is met, the $\mathcal{N}=2$ CP($N-1$) model gives a good description of world-sheet physics. A hierarchy of relevant scales in our theory is displayed in Fig. 5.1.

If we increase μ above the critical value (5.1.47) the non-Abelian strings become thick and their world-sheet dynamics is no longer described by $\mathcal{N}=2$ CP($N-1$) sigma model. The higher-derivative corrections on the world-sheet explode. Note that the physical reason for the growth of the string thickness Δ is the presence of the Higgs branch in $\mathcal{N}=1$ SQCD. Although the classical string solution (4.4.20) retains a finite transverse size, the Higgs branch manifests itself at the quantum level. In particular, the fermion zero modes feel the Higgs branch and acquire long-range logarithmic tails.

Now, let us abstract ourselves from the fact that the theory (5.1.19) is a low-energy effective model on the world sheet of the non-Abelian string. Let us consider

Figure 5.1. Relevant scale hierarchy in the limit $\mu \gg \sqrt{\xi}$.

this model *per se*, with no reference to the underlying four-dimensional theory. Then, of course, the parameter u can be viewed as arbitrary. One can address a subtle question: what happens in the limit $u \to \infty$? In this limit the σ field VEV tends to zero (see Eq. (5.1.39)) and N degenerate vacua coalesce. Moreover, the U(1) gauge field, σ and the fermionic kinks ξ become massless (in addition to the λ_R field which, being the Goldstino in this limit, is necessarily massless). The model seemingly becomes conformal. It is plausible to interpret this conformal fixed point as a phase transition point from the kink deconfinement phase to the Coulomb/confining phase.

A similar phenomenon occurs in two-dimensional conformal $\mathcal{N} = (4, 4)$ supersymmetric gauge theory [195]. In this theory the same tube metric $|d\sigma|^2/|\sigma|^2$ appears (as in (5.1.19), see [192]) and the point $\sigma = 0$ is interpreted as a transition point between two distinct phases.

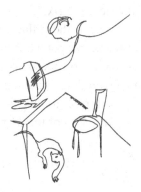

5.2 The M model

In Section 5.1 we learned that the occurrence of the Higgs branch in $\mathcal{N} = 1$ SQCD obscures physics of the non-Abelian strings. Thus, it is highly desirable to get rid of the Higgs branch, keeping $\mathcal{N} = 1$. This was done in [104]. Below we will review key results pertinent to the issue.

To eliminate light states we will introduce a particular $\mathcal{N} = 2$ breaking deformation in the U(N) theory with the potential (5.1.2). Namely, we uplift the quark mass matrix m_A^B (see Eq. (4.1.9) where this matrix is assumed to be diagonal) to the superfield status,

$$m_A^B \to M_A^B \,,$$

and introduce the superpotential

$$\mathcal{W}_M = Q M \tilde{Q} \,. \tag{5.2.1}$$

The matrix M represents N^2 chiral superfields of the mesonic type (they are color singlets). Needless to say, we have to add a kinetic term for M_A^B,

$$S_{M\text{kin}} = \int d^4x \, d^2\theta \, d^2\bar{\theta} \, \frac{2}{h} \, \text{Tr} \, \bar{M}M , \qquad (5.2.2)$$

where h is a new coupling constant (it supplements the set of the gauge couplings). In particular, the kinetic term for the scalar components of M takes the form

$$\int d^4x \left\{ \frac{1}{h} |\partial_\mu M^0|^2 + \frac{1}{h} |\partial_\mu M^a|^2 \right\}, \qquad (5.2.3)$$

where we use the decomposition

$$M_B^A = \frac{1}{2} \delta_B^A \, M^0 + (T^a)_B^A \, M^a . \qquad (5.2.4)$$

At $h = 0$ the matrix field M becomes sterile, it is frozen and in essence returns to the status of a constant numerical matrix. The theory acquires flat directions (a moduli space). With nonvanishing h these flat directions are lifted, and M is determined by the minimum of the scalar potential, see below.

The uplift of the quark mass matrix to superfield is a crucial step which allows us to lift the Higgs branch which would develop in this theory in the large-μ limit if M were a constant matrix. We will refer to this theory as the M model.

The potential $V(q^A, \tilde{q}_A, a^a, a, M^0, M^a)$ of the M model is

$$V(q^A, \tilde{q}_A, a^a, a, M^0, M^a) = \frac{g_2^2}{2} \left(\frac{1}{g_2^2} f^{abc} \bar{a}^b a^c + \text{Tr} \, \bar{q} \, T^a q - \text{Tr} \, \tilde{q} \, T^a \, \bar{\tilde{q}} \right)^2$$

$$+ \frac{g_1^2}{8} \left(\text{Tr} \, \bar{q}q - \text{Tr} \, \tilde{q}\bar{\tilde{q}} - N\xi \right)^2 + \frac{g_2^2}{2} \left| 2\text{Tr} \, \tilde{q} T^a q + \sqrt{2}\mu_2 a^a \right|^2$$

$$+ \frac{g_1^2}{2} \left| \text{Tr} \, \tilde{q}q + \sqrt{N}\mu_1 a \right|^2 + \frac{1}{2} \text{Tr} \left\{ \left| (a + 2\,T^a \, a^a)q + \frac{1}{\sqrt{2}} q(M^0 + 2T^a M^a) \right|^2 \right.$$

$$+ \left. \left| (a + 2\,T^a \, a^a)\bar{\tilde{q}} + \frac{1}{\sqrt{2}} \bar{\tilde{q}}(M^0 + 2T^a M^a) \right|^2 \right\} + \frac{h}{4} |\text{Tr} \, \tilde{q}q|^2 + h|\text{Tr} \, q \, T^a \tilde{q}|^2 .$$

$$(5.2.5)$$

The last two terms here are F terms of the M field. In Eq. (5.2.5) we also introduced the FI D-term for the U(1) field, with the FI parameter ξ.

The FI term triggers the spontaneous breaking of the gauge symmetry. The VEV's of the squark fields and adjoint fields are given by (5.1.3) and (5.1.4), respectively, while the VEV's of M field vanish,

$$\langle M^a \rangle = 0, \quad \langle M^0 \rangle = 0. \qquad (5.2.6)$$

The color-flavor locked form of the quark VEV's in Eq. (5.1.3) and the absence of VEVs of the adjoint field a^a and the meson fields M^a in Eqs. (5.1.4) and (5.2.6) result in the fact that, while the theory is fully Higgsed, a diagonal $SU(N)_{C+F}$ symmetry survives as a global symmetry, much in the same way as in μ-deformations of $\mathcal{N} = 2$ SQCD. Namely, the global rotation

$$q \to UqU^{-1}, \quad a^a T^a \to U a^a T^a U^{-1}, \quad M \to U^{-1} M U, \qquad (5.2.7)$$

is not broken by the VEVs (5.1.3), (5.1.4) and (5.2.6). Here U is a matrix from $SU(N)$. As usual, this symmetry leads to the emergence of orientational zero modes of the Z_N strings in the theory with the potential (5.2.5).

At large μ one can readily integrate out the adjoint fields $\mathcal{A}^a$ and $\mathcal{A}$. The bosonic part of the action of the M model takes the form

$$S = \int d^4x \left\{ \frac{1}{4g_2^2} (F_{\mu\nu}^a)^2 + \frac{1}{4g_1^2} (F_{\mu\nu})^2 + \text{Tr} \, |\nabla_\mu q|^2 + \text{Tr} \, |\nabla_\mu \bar{\tilde{q}}|^2 \right.$$

$$+ \frac{1}{h} |\partial_\mu M^0|^2 + \frac{1}{h} |\partial_\mu M^a|^2 + \frac{g_2^2}{2} \left(\text{Tr} \, \bar{q} \, T^a q - \text{Tr} \, \tilde{q} T^a \, \bar{\tilde{q}} \right)^2$$

$$+ \frac{g_1^2}{8} \left(\text{Tr} \, \bar{q} q - \text{Tr} \, \tilde{q} \bar{\tilde{q}} - N\xi \right)^2 + \text{Tr}|q M|^2 + \text{Tr}|\bar{\tilde{q}} M|^2$$

$$\left. + \frac{h}{4} \left| \text{Tr} \, \tilde{q} q \right|^2 + h \left| \text{Tr} \, q T^a \tilde{q} \right|^2 \right\}. \qquad (5.2.8)$$

The vacuum of this theory is given by Eqs. (5.1.3) and (5.2.6). The mass spectrum of elementary excitations over this vacuum consists of the $\mathcal{N} = 1$ gauge multiplets for the U(1) and SU(N) sectors, with masses given in Eqs. (4.1.16) and (4.1.17). In addition, we have chiral multiplets $\tilde{q}$ and M, with masses

$$m_{U(1)} = \sqrt{\frac{hN\xi}{4}} \qquad (5.2.9)$$

for the U(1) sector, and

$$m_{SU(N)} = \sqrt{\frac{h\xi}{2}} \qquad (5.2.10)$$

for the SU(N) sector.

It is worth emphasizing that there are no massless states in the bulk theory. At $h = 0$ the theory with the potential (5.2.5) develops a Higgs branch in the large-μ limit (see Section 5.1). If $h \neq 0$, M becomes a fully dynamical field. The Higgs branch is lifted, as follows from Eqs. (5.2.9) and (5.2.10).

The $\mathcal{N} = 1$ SQCD with the M field, the M model, belongs to the class of theories introduced by Seiberg [196] to provide a dual description of conventional $\mathcal{N} = 1$ SQCD with the SU(N_c) gauge group and N_f flavors of fundamental matter, where

$$N_c = N_f - N$$

(for reviews see Refs. [197, 198]). There are significant distinctions, however.

Let us outline the main differences of the M model (5.2.8) from those introduced [196] by Seiberg:

(i) The theory (5.2.8) has the U(N) gauge group rather than SU(N);

(ii) It has the FI D term instead of a linear in M superpotential in Seiberg's models;

(iii) Following [104] we consider the case $N_f = N$ which would correspond to Seiberg's $N_c = 0$ in the original SQCD. The theory (5.2.8) is asymptotically free, while Seiberg's dual theories are most meaningful (i.e. have the maximal predictive power with regards to the original strongly coupled $\mathcal{N} = 1$ SQCD) below the left edge of the conformal window, in the range $N_f < (3/2) N_c$, which would correspond to $N_f > 3N$ rather than $N_f = N$. Note that at $N_f > 3N$ the theory (5.2.8) is not asymptotically free and is thus uninteresting from our standpoint.

In addition, it is worth noting that at $N_f > N$ the vacuum (5.1.3), (5.2.6) becomes metastable: supersymmetry is broken [199]. The $N_c = N_f - N$ supersymmetry-preserving vacua have vanishing VEV's of the quark fields and a non-vanishing VEV of the M field.[3] The latter vacua are associated with the gluino condensation in pure SU(N) theory, $\langle \lambda\lambda \rangle \neq 0$, arising upon decoupling N_f flavors [197]. In the case $N_f = N$ to which we limit ourselves the vacuum (5.1.3), (5.2.6) preserves supersymmetry. Thus, despite a conceptual similarity between Seiberg's models and ours, dynamical details are radically different.

Now, it is time to pass to solutions for non-Abelian BPS strings in the M model [104]. Much in the same way as in Section 5.1 we use the *ansatz* (5.1.16). Moreover, we set the adjoint fields and the M fields to zero. With these simplifications the $\mathcal{N} = 1$ model with the potential (5.2.5) reduces to the model (4.2.2) which we used previously in the original construction of the non-Abelian strings.

In particular, the solution for the elementary string is given by (4.4.4). Moreover, the bosonic part of the effective world-sheet theory is again described by the CP($N-1$) sigma model (4.4.9) with the coupling constant β determined by (4.4.15). The scale of this CP($N-1$) model is given by Eq. (5.1.18) in the limit of large μ.

The full construction of the world-sheet theory in the M model has not been yet carried out. One can conjecture as to what the fermion part of this theory is. There are good reasons to expect that we will get the heterotic $\mathcal{N} = (0,2)$ CP($N-1$)

---

[3] This is correct for the version of the theory with the ξ-parameter introduced via superpotential.

theory much in the same way as in Section 5.1.2 (see also Appendix B.5). The relation between the bulk and world sheet deformation parameters are likely to change, but all consequences (such as spontaneous SUSY breaking at the quantum level) presumably will stay intact.

To conclude this section let us note a somewhat related development: *non*-BPS non-Abelian strings were considered in metastable vacua of a dual description of $\mathcal{N} = 1$ SQCD at $N_f > N$ in Ref. [200].

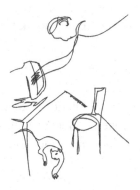

5.3 Confined non-Abelian monopoles

As was mentioned, the effective low-energy Lagrangian describing world-sheet physics of the non-Abelian string in the M model, must be supersymmetric, presumably, $\mathcal{N} = (0, 2)$. The heterotic sigma model dynamics is known (see Section 5.1.3); in particular, we will have N degenerate vacua and kinks that interpolate between them, similar to the kinks that emerge in $\mathcal{N} = 2$ SQCD. These kinks are interpreted as (confined) non-Abelian monopoles [165, 132, 133], the descendants of the 't Hooft–Polyakov monopole.

Let us discuss what happens with these monopoles as we deform our theory and eventually end up with the M model. It is convenient to split this deformation into several distinct stages. We will describe what happens with the monopoles as one passes from one stage to another. Some of these steps involving deformations of $\mathcal{N} = 2$ SQCD were already discussed in Section 4.5. Here we focus on deformations of $\mathcal{N} = 2$ SQCD leading to the M model.

A qualitative evolution of the monopoles under consideration as a function of the relevant parameters is presented in Fig. 4.3.

(i) We start from $\mathcal{N} = 2$ SQCD turning off the $\mathcal{N} = 2$ breaking parameters h and μ's as well as the FI parameter in the potential (5.2.5), i.e. we start from the Coulomb branch of the theory,

$$\mu_1 = \mu_2 = 0, \quad h = 0, \quad \xi = 0, \quad M \neq 0. \qquad (5.3.1)$$

As was explained in Section 5.2, the field M is frozen in this limit and can take arbitrary values (the notorious flat direction). The matrix M_B^A plays the role of a fixed mass matrix for the quark fields. As a first step let us consider the diagonal matrix M, with distinct diagonal entries,

$$M_B^A = \text{diag}\,\{M_1, \dots, M_N\}. \tag{5.3.2}$$

Shifting the field a one can always make $\sum_A M_A = 0$ in the limit $\mu_1 = 0$. Therefore $M^0 = 0$. If all M_A's are different the gauge group SU(N) is broken down to U(1)$^{(N-1)}$ by a VEV of the SU(N) adjoint scalar (see (4.1.11)),

$$\langle a_l^k \rangle = -\frac{1}{\sqrt{2}} \delta_l^k M_l. \tag{5.3.3}$$

Thus, as was already discussed in Section 4.5, there are 't Hooft–Polyakov monopoles embedded in the broken gauge SU(N). Classically, on the Coulomb branch the masses of $(N-1)$ elementary monopoles are proportional to

$$|(M_A - M_{A+1})|/g_2^2. $$

In the limit $(M_A - M_{A+1}) \to 0$ the monopoles tend to become massless, formally, in the classical approximation. Simultaneously their size becomes infinite [112]. The mass and size are stabilized by highly quantum confinement effects. The monopole confinement occurs in the Higgs phase, at $\xi \neq 0$.

(ii) Now let us make the FI parameter ξ non-vanishing. This triggers the squark condensation. The theory is in the Higgs phase. We still keep $\mathcal{N} = 2$ breaking parameters h and μ's vanishing,

$$\mu_1 = \mu_2 = 0, \quad h = 0, \quad \xi \neq 0, \quad M \neq 0. \tag{5.3.4}$$

The squark condensation leads to formation of the Z_N strings. Monopoles become confined by these strings. As we discussed in Section 4.5, $(N-1)$ elementary monopoles become junctions of pairs of elementary strings.

Now, if we reduce $|\Delta M_A|$,

$$\Lambda_{\text{CP}(N-1)} \ll |\Delta M_A| \ll \sqrt{\xi}, \tag{5.3.5}$$

the size of the monopole along the string $\sim |(M_A - M_{A+1})|^{-1}$ becomes larger than the transverse size of the attached strings. The monopole becomes a *bona fide* confined monopole (the lower left corner of Fig. 4.3). At nonvanishing ΔM_A the effective theory on the string world sheet is the CP($N-1$) model with twisted mass terms [165, 132, 133], see Section 4.4.4. Two Z_N strings attached to an elementary monopole correspond to two "neighboring" vacua of the CP($N - 1$) model. The monopole (a.k.a. the string junction of two Z_N strings) manifests itself as a kink interpolating between these two vacua.

(iii) Next, we switch off the mass differences ΔM_A still keeping the $\mathcal{N} = 2$ breaking parameters vanishing,

$$\mu_1 = \mu_2 = 0, \quad h = 0, \quad \xi \neq 0, \quad M = 0. \qquad (5.3.6)$$

The values of the twisted masses in CP($N - 1$) model coincide with ΔM_A while the size of the twisted-mass sigma-model kink/confined monopole is of the order of $\sim |(M_A - M_{A+1})|^{-1}$.

As we decrease ΔM_A approaching $\Lambda_{\mathrm{CP}(N-1)}$ and then getting below the scale $\Lambda_{\mathrm{CP}(N-1)}$, the monopole size grows, and, classically, it would explode. This is where quantum effects in the world sheet theory take over. It is natural to refer to this domain of parameters as the "regime of highly quantum dynamics." While the thickness of the string (in the transverse direction) is $\sim \xi^{-1/2}$, the z-direction size of the kink representing the confined monopole in the highly quantum regime is much larger, $\sim \Lambda_{\mathrm{CP}(N-1)}^{-1}$, see the lower right corner in Fig. 4.3.

In this regime the confined monopoles become non-Abelian. They no longer carry average magnetic flux since

$$\langle n^l \rangle = 0, \qquad (5.3.7)$$

in the strong coupling limit of the CP($N - 1$) model [159]. The kink/monopole belongs to the fundamental representation of the SU(N)$_{C+F}$ group [159, 120].

(iv) Thus, with vanishing ΔM_A we still have confined "monopoles" (interpreted as kinks) stabilized by quantum effects in the world sheet CP($N - 1$) model. Now we can finally switch on the $\mathcal{N} = 2$ breaking parameters μ_i and h,

$$\mu_i \neq 0, \quad h \neq 0, \quad \xi \neq 0, \quad M = 0. \qquad (5.3.8)$$

Note that the last equality here is automatically satisfied in the vacuum, see Eq. (5.2.6).

As was discussed in Section 5.2 the effective world sheet description of the non-Abelian string is given by a heterotic deformation of the supersymmetric CP($N - 1$) model. This two-dimensional theory has N vacua which should be interpreted as N elementary non-Abelian strings in the quantum regime, with kinks interpolating between these vacua. These kinks should be interpreted as non-Abelian confined monopoles/string junctions.

Note that although the adjoint fields are still present in the theory (5.2.5) their VEV's vanish (see (5.1.4)) and the monopoles cannot be seen in the semiclassical approximation. They are seen solely as world sheet kinks. Their mass and inverse size are determined by Λ_σ which in the limit of large μ_i is given by Eq. (5.1.18).

(v) At the last stage, we take the limit of large masses of the adjoint fields in order to eliminate them from the physical spectrum altogether,

$$\mu_i \to \infty, \qquad h \neq 0, \qquad \xi \neq 0, \qquad M = 0. \tag{5.3.9}$$

The theory flows to $\mathcal{N} = 1$ SQCD extended by the M field.

In this limit we get a remarkable result: although the adjoint fields are eliminated from our theory and the monopoles cannot be seen in any semiclassical description, our analysis shows that confined non-Abelian monopoles still exist in the theory (5.2.8). They are seen as kinks in the effective world sheet theory on the non-Abelian string.

(vi) The confined monopoles are in the highly quantum regime, so they carry no average magnetic flux (see Eq. (5.3.7)). Thus, they are non-Abelian. Moreover, they acquire global flavor quantum numbers. In fact, they belong to the fundamental representation of the global SU(N)$_{C+F}$ group (see Refs. [159, 120] where this phenomenon is discussed in the context of the CP($N - 1$)-model kinks).

It is quite plausible that the emergence of these non-Abelian monopoles can shed light on mysterious objects introduced by Seiberg: "dual magnetic" quarks which play an important role in the description of $\mathcal{N} = 1$ SQCD at strong coupling [196, 197].

5.4 Index theorem

In this section we will discuss an index theorem establishing the number of the fermion zero modes on the string. For definiteness we will consider the M model [104]. Similar theorems can be easily proved for ordinary $\mathcal{N} = 1$ SQCD (5.1.13) as well as for theories with potentials (5.1.2) or (5.2.5) at intermediate values of μ. They generalize index theorems obtained long ago for simple non-supersymmetric models [201].

The fermionic part of the action of the model (5.2.8) is

$$
\begin{aligned}
S_{\text{ferm}} = \int d^4x \bigg\{ &\frac{i}{g_2^2}\bar{\lambda}^a \slashed{D}\lambda^a + \frac{i}{g_1^2}\bar{\lambda}\slashed{\partial}\lambda + \text{Tr}[\bar{\psi}i\slashed{\nabla}\psi] + \text{Tr}[\tilde{\psi}i\slashed{\nabla}\bar{\tilde{\psi}}] \\
&+ \frac{2i}{h}\text{Tr}[\bar{\zeta}\slashed{\partial}\zeta] + \frac{i}{\sqrt{2}}\text{Tr}[\bar{q}(\lambda\psi) - (\bar{\tilde{\psi}}\lambda)\bar{\tilde{q}} + (\tilde{\psi}\bar{\lambda})q - \tilde{q}(\bar{\lambda}\bar{\psi})] \\
&+ \frac{i}{\sqrt{2}}\text{Tr}[\bar{q}\,2T^a\,(\lambda^a\psi) - (\bar{\tilde{\psi}}\lambda^a)\,2T^a\,\bar{\tilde{q}} + (\tilde{\psi}\bar{\lambda}^a)\,2T^a\,q - \tilde{q}\,2T^a\,(\bar{\lambda}^a\bar{\psi})] \\
&+ i\,\text{Tr}[\bar{q}(\psi\zeta) + (\tilde{\psi}q\zeta) + (\bar{\tilde{\psi}}\bar{\tilde{q}}\bar{\zeta}) + \bar{q}(\bar{\psi}\bar{\zeta})] \\
&+ i\,\text{Tr}(\bar{\tilde{\psi}}\psi M + \tilde{\psi}\bar{\psi}\bar{M}) \bigg\},
\end{aligned}
\tag{5.4.1}
$$

where the matrix color-flavor notation is used for the matter fermions $(\psi^\alpha)^{kA}$ and $(\tilde{\psi}^\alpha)_{Ak}$, and the traces are performed over the color-flavor indices. Moreover, ζ denotes the fermion component of the matrix M superfield,

$$
\zeta_B^A = \frac{1}{2}\delta_B^A\,\zeta^0 + (T^a)_B^A\,\zeta^a.
\tag{5.4.2}
$$

In order to find the number of the fermion zero modes in the background of the non-Abelian string solution (4.4.4) we have to carry out the following program. Since our string solution depends only on two coordinates x_i ($i = 1, 2$), we can reduce our theory to two dimensions. Given the theory defined on the (x_1, x_2) plane we have to identify an axial current and derive the anomalous divergence for this current. In two dimensions the axial current anomaly takes the form

$$
\partial_i j_{i5} \sim F^*,
\tag{5.4.3}
$$

where $F^* = (1/2)\varepsilon_{ij}F_{ij}$ is the dual U(1) field strength in two dimensions.

Then, the integral over the left-hand side over the (x_1, x_2) plane gives us the index of the 2D Dirac operator ν coinciding with the number of the 2D left-handed minus 2D right-handed zero modes of this operator in the given background field. The integral over the right-hand side is proportional to the string flux. This will fix the number of the chiral fermion zero modes[4] on the string with the given flux. Note that the reduction of the theory to two dimensions is an important step in this program. The anomaly relation in four dimensions involves the instanton charge F^*F rather than the string flux and is therefore useless for our purposes.

---

[4] Chirality is understood here as the two-dimensional chirality.

Table 5.1. *The* $\mathrm{U}(1)_R$ *and* $\mathrm{U}(1)_{\tilde{R}}$ *charges of fields of the two-dimensional reduction of the theory.*

Field	ψ_+	ψ_-	$\tilde{\psi}_+$	$\tilde{\psi}_-$	λ_+	λ_-	ζ_+	ζ_-	q	$\tilde{q}$
$\mathrm{U}(1)_R$ charge	-1	1	-1	1	-1	1	-1	1	0	0
$\mathrm{U}(1)_{\tilde{R}}$ charge	-1	1	1	-1	-1	1	1	-1	0	0

The reduction of $\mathcal{N} = 1$ gauge theories to two dimensions is discussed in detail in [157] and here we will be brief. Following [157] we use the rules

$$\psi^\alpha \to (\psi^-, \psi^+), \quad \tilde{\psi}^\alpha \to (\tilde{\psi}^-, \tilde{\psi}^+),$$
$$\lambda^\alpha \to (\lambda^-, \lambda^+), \quad \zeta^\alpha \to (\zeta^-, \zeta^+). \tag{5.4.4}$$

With these rules the Yukawa interactions in (4.4.22) take the form

$$\mathcal{L}_{\text{Yukawa}} = i\sqrt{2}\,\mathrm{Tr}\left[-\bar{q}(\hat{\lambda}_-\psi_+ - \hat{\lambda}_+\psi_-) + (\tilde{\psi}_-\hat{\lambda}_+ - \tilde{\psi}_+\hat{\lambda}_-)\bar{\tilde{q}} + \text{H.c.}\right]$$
$$- i\,\mathrm{Tr}\left[\tilde{q}(\psi_-\zeta_+ - \psi_+\zeta_-) + (\tilde{\psi}_-q\zeta_+ - \tilde{\psi}_+q\zeta_-) + \text{H.c.}\right], \tag{5.4.5}$$

where the color matrix $\hat{\lambda} = (1/2)\,\lambda + T^a\lambda^a$.

It is easy to see that $\mathcal{L}_{\text{Yukawa}}$ is classically invariant under the chiral $\mathrm{U}(1)_R$ transformations with the $\mathrm{U}(1)_R$ charges presented in Table 5.1. The axial current associated with this $\mathrm{U}(1)_R$ is not anomalous [157]. This is easy to understand. In two dimensions the chiral anomaly comes from the diagram shown in Fig. 5.2. The $\mathrm{U}(1)_R$ chiral charges of the fields ψ and $\tilde{\psi}$ are the same while their electric charges are opposite. This leads to cancellation of their contributions to this diagram.

It turns out that for the particular string solution we are interested in the classical two-dimensional action has more symmetries than generically, for a general background. To see this, please note that the field $\tilde{q}$ vanishes on the string solution (4.4.4), see (5.1.16). Then the Yukawa interactions (5.4.5) reduce to

$$i\sqrt{2}\,\mathrm{Tr}\left[-\bar{q}(\hat{\lambda}_-\psi_+ - \hat{\lambda}_+\psi_-)\right] - i\,\mathrm{Tr}\left[\tilde{\psi}_-q\zeta_+ - \tilde{\psi}_+q\zeta_-\right] + \text{H.c.} \tag{5.4.6}$$

The fermion ψ interacts only with λ's while the fermion $\tilde{\psi}$ interacts only with ζ. Note also that the interaction in the last line in (5.4.1) is absent because $M = 0$ on the string solution. This property allows us to introduce another chiral symmetry in the theory, the one which is relevant for the string solution. We will refer to this extra chiral symmetry as $\mathrm{U}(1)_{\tilde{R}}$.

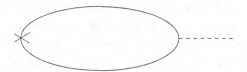

Figure 5.2. Diagram for the chiral anomaly in two dimensions. The solid lines denote fermions ψ, $\tilde{\psi}$, the dashed line denotes the photon, while the cross denotes insertion of the axial current.

The $U(1)_{\tilde{R}}$ charges of our set of fields are also shown in Table 5.1. Note that ψ and $\tilde{\psi}$ have the opposite charges under this symmetry. The corresponding current then has the form

$$\tilde{j}_{i5} = \begin{pmatrix} \bar{\psi}_-\psi_- - \bar{\psi}_+\psi_+ - \bar{\tilde{\psi}}_-\tilde{\psi}_- + \bar{\tilde{\psi}}_+\tilde{\psi}_+ + \cdots \\ -i\bar{\psi}_-\psi_- - i\bar{\psi}_+\psi_+ + i\bar{\tilde{\psi}}_-\tilde{\psi}_- + i\bar{\tilde{\psi}}_+\tilde{\psi}_+ + \cdots \end{pmatrix}, \qquad (5.4.7)$$

where the ellipses stand for terms associated with the λ and ζ fields which do not contribute to the anomaly relation.

It is clear that the $U(1)_{\tilde{R}}$ symmetry is anomalous in quantum theory. The contributions of the fermions ψ and $\tilde{\psi}$ double in the diagram in Fig. 5.2 rather than cancel. It is not difficult to find the coefficient in the anomaly formula

$$\partial_i \tilde{j}_{i5} = \frac{N^2}{\pi} F^*, \qquad (5.4.8)$$

which can be normalized e.g. from [202]. The factor N^2 appears due to the presence of $2N^2$ fermions ψ^{kA} and $\tilde{\psi}_{Ak}$.

Now, taking into account the fact that the flux of the Z_N string under consideration is

$$\int d^2x \, F^* = \frac{4\pi}{N}, \qquad (5.4.9)$$

(see the expression for the $U(1)$ gauge field for the solution (4.2.6) or (4.4.4)) we conclude that the total number of the fermion zero modes in the string background[5]

$$\nu = 4N. \qquad (5.4.10)$$

This number can be decomposed as

$$\nu = 4N = 4(N-1) + 4, \qquad (5.4.11)$$

---

[5] Equations (5.4.9) and (5.4.10) are very similar in essence to analogous four-dimensional relations connecting the instanton topological charge with the number of the fermion zero modes in the instanton background. For a review see [203].

where 4 is the number of the supertranslational modes while $4(N-1)$ is the number of the superorientational modes. Four supertranslational modes are associated with four fermion fields in the two-dimensional effective theory on the string world sheet, which are superpartners of the bosonic translational moduli x_0 and y_0. Furthermore, $4(N-1)$ corresponds to $4(N-1)$ fermion fields in the $\mathcal{N}=2$ CP$(N-1)$ model on the string world sheet. CP$(N-1)$ describes dynamics of the orientational moduli of the string. For $N=2$ the latter number ($4(N-1)=4$) counts four fermion fields χ_1^a, χ_2^a in the model (4.4.31).

A similar theorem can be formulated for $\mathcal{N}=1$ theory with the potential (5.1.2) as well; it implies $4(N-1)$ orientational zero modes in this case too, i.e. the doubling of the number of the fermion zero modes on the string as compared with the one which follows from "BPS-ness."

In [189] and [104] four orientational fermion zero modes were found explicitly in $\mathcal{N}=1$ SQCD and the M model, by solving the Dirac equations in the string background. Note that these fermion zero modes in the M model are perfectly normalizable provided we keep the coupling constant h non-vanishing. Instead, in conventional $\mathcal{N}=1$ SQCD without the M field the second pair of the fermion zero modes (proportional to χ_1^a) become non-normalizable in the large-μ limit [189]. This is related to the presence of the Higgs branch and massless bulk states in conventional $\mathcal{N}=1$ SQCD. As was already mentioned more than once, in the M model, Eq. (5.2.8), we have no massless states in the bulk.

Note that in both translational and orientational sectors the number of the fermion zero modes is twice larger than the one dictated by 1/2 "BPS-ness." Fermion supertranslational zero modes of the non-Abelian string in $\mathcal{N}=1$ theory with the potential (5.1.2) were found in [191]. Just like superorientational modes, they acquire long-range tails in the large-μ limit and become non-normalizable.

6

Non-BPS non-Abelian strings

In this chapter we will review non-BPS non-Abelian strings. In particular, they appear in non-supersymmetric theories. We will see that, although for BPS strings in supersymmetric theories the transition from quasiclassical to quantum regimes in the world sheet theory on the string goes smoothly (see Section 4.4.4), for the non-Abelian strings in non-supersymmetric theories these two regimes are separated by a phase transition.

Next, we will show that the same behavior is typical for non-BPS strings in supersymmetric gauge theories. As an example we consider non-Abelian strings in the so-called $\mathcal{N} = 1^*$ theory which is a deformed $\mathcal{N} = 4$ supersymmetric theory with supersymmetry broken down to $\mathcal{N} = 1$ in a special way.

6.1 Non-Abelian strings in non-supersymmetric theories

In this section we will review some results reported in [154, 164] treating non-Abelian strings in non-supersymmetric gauge theories. The theory studied in [154] is essentially a bosonic part of $\mathcal{N} = 2$ supersymmetric QCD with the gauge group $SU(N) \times U(1)$ described in Chapter 4 in the supersymmetric setting.[1] The action of this model is

$$
S = \int d^4x \left\{ \frac{1}{4g_2^2} \left(F_{\mu\nu}^a \right)^2 + \frac{1}{4g_1^2} \left(F_{\mu\nu} \right)^2 + \frac{1}{g_2^2} |D_\mu a^a|^2 \right.
$$
$$
+ |\nabla^\mu \varphi^A|^2 + \frac{g_2^2}{2} \left(\bar{\varphi}_A T^a \varphi^A \right)^2 + \frac{g_1^2}{8} \left(|\varphi^A|^2 - N\xi \right)^2
$$
$$
+ \left. \frac{1}{2} \left| \left(a^a T^a + \sqrt{2} m_A \right) \varphi^A \right|^2 + \frac{i\theta}{32\pi^2} F_{\mu\nu}^a F_{\mu\nu}^{*a} \right\}, \quad (6.1.1)
$$

---

[1] In addition to the substitution (4.2.1) we discard the $f^{abc} \bar{a}^b a^c$ term in Eq. (4.1.9). This term plays no role in the consideration presented below.

where $F^{*a}_{\mu\nu} = (1/2)\,\varepsilon_{\mu\nu\alpha\beta}F_{\alpha\beta}$ and θ is the vacuum angle. This model is a bosonic part of the $\mathcal{N} = 2$ supersymmetric theory (4.1.7) where, instead of two squark fields q^k and $\tilde{q}_k$, only one fundamental scalar φ^k is introduced for each flavor $A = 1,\ldots,N_f$, see the reduced model (4.2.2) in Section 4.2. We also limit ourselves to the case $N_f = N$ and drop the neutral scalar field a present in (4.1.7) as it plays no role in the string solutions. To keep the theory at weak coupling we consider large values of the parameter ξ in (6.1.1), $\xi \gg \Lambda_{\mathrm{SU}(N)}$.

We assume here that

$$\sum_{A=1}^{N} m_A = 0. \tag{6.1.2}$$

Later on it will be convenient to make a specific choice of the parameters m_A, namely,

$$m^A = m \times \mathrm{diag}\{e^{2\pi i/N},\, e^{4\pi i/N},\ldots,e^{2(N-1)\pi i/N},\, 1\}, \tag{6.1.3}$$

where m is a single common parameter. Then the constraint (6.1.2) is automatically satisfied. We can (and will) assume m to be real and positive. We also introduce a θ term in the model (6.1.1).

Clearly the vacuum structure of the model (6.1.1) is the same as of the theory (4.1.7), see Section 4.1. Moreover, the Z_N string solutions are the same; they are given in Eq. (4.2.6). The adjoint field plays no role in this solution and is given by its VEV (4.1.11). The tensions of these strings are given classically by Eq. (4.2.12). However, in contrast with the supersymmetric theory, now the tensions of Z_N strings acquire quantum corrections in loops.

If masses of the fundamental matter vanish in (6.1.1) this theory has unbroken $\mathrm{SU}(N)_{C+F}$ much in the same way as the theory (4.1.7). In this limit the Z_N strings acquire orientational zero modes and become non-Abelian. The corresponding solution for the elementary non-Abelian string is given by Eq. (4.3.1). Below we will consider two-dimensional effective low-energy theory on the world sheet of such non-Abelian string. Its physics appears to be quite different as compared with the one in the supersymmetric case.

6.1.1 World-sheet theory

Derivation of the effective world-sheet theory for the non-Abelian string in the model (6.1.1) can be carried out much in the same way as in the supersymmetric case [154], see Section 4.4. The world-sheet theory now is two-dimensional non-supersymmetric CP($N - 1$) model (4.4.9). Its coupling constant β is given by the coupling constant g_2^2 of the bulk theory via the relation (4.4.10). Classically the normalization integral I is given by (4.4.11). Then it follows that $I = 1$ as in supersymmetric case. However, now we expect quantum corrections to modify this result. In particular, I can become a function of N in quantum theory.

Now, let us discuss the impact of the θ term which we introduced in our bulk theory (6.1.1). At first sight, seemingly it cannot produce any effect because our string is magnetic. However, if one allows for slow variations of n^l in z and t, one immediately observes that the electric field is generated via $A_{0,3}$ in Eq. (4.4.5). Substituting F_{ki} from (4.4.7) into the θ term in the action (6.1.1) and taking into account the contribution from F_{kn} times F_{ij} ($k, n = 0, 3$ and $i, j = 1, 2$) we get the topological term in the effective CP($N - 1$) model (4.4.9) in the form

$$S^{(1+1)} = \int dt\, dz \left\{ 2\beta \left[(\partial_\alpha n^* \partial_\alpha n) + (n^* \partial_\alpha n)^2 \right] - \frac{\theta}{2\pi} I_\theta \varepsilon_{\alpha\gamma} (\partial_\alpha n^* \partial_\gamma n) \right\},$$

(6.1.4)

where I_θ is another normalizing integral given by the formula

$$
\begin{aligned}
I_\theta &= - \int dr \left\{ 2 f_{NA} (1 - \rho) \frac{d\rho}{dr} + (2\rho - \rho^2) \frac{df}{dr} \right\} \\
&= \int dr \frac{d}{dr} \{ 2 f_{NA} \rho - \rho^2 f_{NA} \}.
\end{aligned}
$$

(6.1.5)

As is clearly seen, the integrand here reduces to a total derivative, and is determined by the boundary conditions for the profile functions ρ and f_{NA}. Substituting (4.4.6), (4.4.8), and (4.2.8), (4.2.7) we get

$$I_\theta = 1,$$

(6.1.6)

independently of the form of the profile functions. This latter circumstance is perfectly natural for the topological term.

The additional term in the CP($N - 1$) model (6.1.4) we have just derived is the θ term in the standard normalization. The result (6.1.6) could have been expected since physics is 2π-periodic with respect to θ both in the four-dimensional bulk gauge theory and in the two-dimensional world sheet CP($N - 1$) model. The result (6.1.6) is not sensitive to the presence of supersymmetry. It will hold in supersymmetric models as well. Note that the complexified bulk coupling constant converts into the complexified world sheet coupling constant,

$$\tau = \frac{4\pi}{g_2^2} + i \frac{\theta}{2\pi} \rightarrow 2\beta + i \frac{\theta}{2\pi}.$$

(6.1.7)

Now let us introduce small masses for the fundamental matter in (6.1.1). Clearly the diagonal color-flavor group SU(N)$_{C+F}$ is now broken by adjoint VEV's down to $U(1)^{N-1} \times Z_N$. Still, the solutions for the Abelian (or Z_N) strings are the same as was discussed in Section 4.4.4 since the adjoint field does not enter these solutions. In particular, we have N distinct Z_N string solutions depending on what particular squark winds at infinity, see Section 4.4.4. Say, the string solution with the winding last flavor is still given by Eq. (4.2.6).

What is changed with the color-flavor $SU(N)_{C+F}$ explicitly broken by $m_A \neq 0$, is that the rotations (4.3.1) no longer generate zero modes. In other words, the fields n^ℓ become quasimoduli: a shallow potential (4.4.49) for the quasi-moduli n^l on the string world sheet is generated [132, 133, 154]. Note that we can replace $\tilde{m}_A$ by m_A due to the condition (6.1.2). This potential is shallow as long as $m_A \ll \sqrt{\xi}$.

The potential simplifies if the mass terms are chosen according to (6.1.3),

$$V_{\mathrm{CP}(N-1)} = 2\beta\, m^2 \left\{ 1 - \left| \sum_{\ell=1}^{N} e^{2\pi i\, \ell/N}\, |n^\ell|^2 \right|^2 \right\}. \qquad (6.1.8)$$

This potential is obviously invariant under the cyclic Z_N substitution

$$\ell \to \ell + k, \quad n^\ell \to n^{\ell+k}, \quad \forall \ell, \qquad (6.1.9)$$

with k fixed. This property will be exploited below.

Now our effective two-dimensional theory on the string world sheet becomes a massive $\mathrm{CP}(N-1)$ model (see Appendix B). As in the supersymmetric case the potential (6.1.8) has N vacua at

$$n^\ell = \delta^{\ell\ell_0}, \quad \ell_0 = 1, 2, \ldots, N. \qquad (6.1.10)$$

These vacua correspond to N distinct Abelian Z_N strings with $\varphi^{\ell_0\ell_0}$ winding at infinity, see Eq. (4.4.4).

6.1.2 Physics in the large-N limit

The massless non-supersymmetric $\mathrm{CP}(N-1)$ model (6.1.4) was solved a long time ago by Witten in the large-N limit [159]. The massive case with the potential (6.1.8) was considered at large N in [154, 164] in connection with the non-Abelian strings. Here we will briefly review this analysis.

As was discussed in Section 4.4.4, the $\mathrm{CP}(N-1)$ model can be understood as a strong coupling limit of a $U(1)$ gauge theory. The action has the form

$$S = \int d^2x \left\{ 2\beta\, |\nabla_k n^\ell|^2 + \frac{1}{4e^2} F_{kp}^2 + \frac{1}{e^2} |\partial_k \sigma|^2 - \frac{\theta}{2\pi} \varepsilon_{kp}\, \partial_k A_p \right.$$
$$\left. + 4\beta \left| \sigma - \frac{\tilde{m}_\ell}{\sqrt{2}} \right|^2 |n^\ell|^2 + 2e^2 \beta^2 (|n^\ell|^2 - 1)^2 \right\}, \qquad (6.1.11)$$

where we also included the θ term. As in the supersymmetric case, in the limit $e^2 \to \infty$ the σ field can be eliminated via the algebraic equation of motion which leads to the theory (6.1.4) with the potential (4.4.49).

The Z_N-cyclic symmetry (6.1.9) now takes the form

$$\sigma \to e^{i\frac{2\pi k}{N}}\sigma, \quad n^\ell \to n^{\ell+k}, \quad \forall \ell, \tag{6.1.12}$$

where k is fixed.

It turns out that the non-supersymmetric version of the massive CP($N-1$) model (6.1.11) has two phases separated by a phase transition [154, 164]. At large values of the mass parameter m we have the Higgs phase while at small m the theory is in the Coulomb/confining phase.

The Higgs phase

At large m, $m \gg \Lambda_\sigma$, the renormalization group flow of the coupling constant β in (6.1.11) is frozen at the scale m. Thus, the model at hand is at weak coupling and the quasiclassical analysis is applicable. The potential (6.1.8) has N degenerate vacua which are labeled by the order parameter $\langle\sigma\rangle$, the vacuum configuration being

$$n^\ell = \delta^{\ell\ell_0}, \quad \sigma = \frac{\tilde{m}_{\ell_0}}{\sqrt{2}}, \quad \ell_0 = 1,\ldots,N, \tag{6.1.13}$$

as in the supersymmetric case, see (4.4.53). In each given vacuum the Z_N symmetry (6.1.12) is spontaneously broken.

These vacua correspond to Abelian Z_N strings of the bulk theory. N vacua of the world-sheet theory have strictly degenerate vacuum energies. From the four-dimensional point of view this means that we have N strictly degenerate Z_N strings.

There are $2(N-1)$ elementary excitations. Here we count real degrees of freedom. The action (6.1.11) contains N complex fields n^ℓ. The common phase of n^{ℓ_0} is gauged away. The condition $|n^\ell|^2 = 1$ eliminates one more field. These elementary excitations have physical masses

$$M_\ell = |m_\ell - m_{\ell_0}|, \quad \ell \neq \ell_0. \tag{6.1.14}$$

Besides, there are kinks (domain "walls" which are particles in two dimensions) interpolating between these vacua. Their masses scale as

$$M_\ell^{\text{kink}} \sim \beta M_\ell. \tag{6.1.15}$$

The kinks are much heavier than elementary excitations at weak coupling. Note that they have nothing to do with Witten's n solitons [159] identified as solitons at strong coupling. The point of phase transition separates these two classes of solitons.

As was already discussed in the supersymmetric case (see Section 4.5) the flux of the Abelian 't Hooft–Polyakov monopole is the difference of the fluxes of two "neighboring" strings, see (4.5.1). Therefore, the confined monopole in this regime is obviously a junction of two distinct Z_N strings. It is seen as a quasiclassical kink interpolating between the "neighboring" ℓ_0th and $(\ell_0 + 1)$th vacua of the effective massive $CP(N - 1)$ model on the string world sheet. A monopole can move freely along the string as both attached strings are tension-degenerate.

The Coulomb/confining phase

Now let us discuss the Coulomb/confining phase of the theory occurring at small m. As was mentioned, at $m = 0$ the $CP(N - 1)$ model was solved by Witten in the large-N limit [159]. The model at small m is very similar to Witten's solution. (In fact, in the large-N limit it is just the same.) The paper [164] presents a generalization of Witten's analysis to the massive case which is then used to study the phase transition between the Z_N asymmetric and symmetric phases. Here we will briefly summarize Witten's results for the massless model.

The non-supersymmetric $CP(N - 1)$ model is asymptotically free (as its super-symmetric version) and develops its own scale Λ_σ. If $m = 0$, classically the field n^ℓ can have arbitrary direction; therefore, one might naively expect spontaneous breaking of $SU(N)$ and the occurrence of massless Goldstone modes. This cannot happen in two dimensions. Quantum effects restore the full symmetry making the vacuum unique. Moreover, the condition $|n^\ell|^2 = 1$ gets in effect relaxed. Due to strong coupling we have more degrees of freedom than in the original Lagrangian, namely all N fields n become dynamical and acquire masses Λ_σ.

This is not the end of the story, however. In addition, one gets another composite degree of freedom. The $U(1)$ gauge field A_k acquires a standard kinetic term at one-loop level,[2] of the form

$$N \Lambda^{-2} F_{kp} F_{kp}. \tag{6.1.16}$$

Comparing Eq. (6.1.16) with (6.1.11) we see that the charge of the n fields with respect to this photon is $1/\sqrt{N}$. The Coulomb potential between two charges in two dimensions is linear in separation between these charges. The linear potential scales as

$$V(R) \sim \frac{\Lambda_\sigma^2}{N} R, \tag{6.1.17}$$

where R is separation. The force is attractive for pairs $\bar{n}$ and n, leading to formation of weakly coupled bound states (weak coupling is the manifestation of the $1/N$

---

[2] By loops here we mean perturbative expansion in $1/N$ perturbation theory.

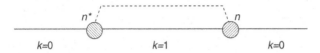

Figure 6.1. Linear confinement of the n-n^* pair. The solid straight line represents the string. The dashed line shows the vacuum energy density (normalizing $\mathcal{E}_0$ to zero).

suppression of the confining potential). Charged states are eliminated from the spectrum. This is the reason why the n fields were called "quarks" by Witten. The spectrum of the theory consists of $\bar{n}n$- "mesons." The picture of confinement of n's is shown in Fig. 6.1.

The validity of the above consideration rests on large N. If N is not large Witten's solution [159] ceases to be applicable. It remains valid in the qualitative sense, however. Indeed, at $N = 2$ the model was solved exactly [204, 205] (see also [206]). Zamolodchikovs found that the spectrum of the O(3) model consists of a triplet of degenerate states (with mass $\sim \Lambda_\sigma$). At $N = 2$ the action (6.1.11) is built of doublets. In this sense one can say that Zamolodchikovs' solution exhibits confinement of doublets. This is in qualitative accord with the large-N solution [159].

Inside the $\bar{n}\,n$ mesons, we have a constant electric field, see Fig. 6.1. Therefore the spatial interval between $\bar{n}$ and n has a higher energy density than the domains outside the meson.

Modern understanding of the vacuum structure of the massless CP($N-1$) model [207] (see also [208]) allows one to reinterpret confining dynamics of the n fields in different terms [155, 154]. Indeed, at large N, along with the unique ground state, the model has $\sim N$ quasi-stable local minima, quasi-vacua, which become absolutely stable at $N = \infty$. The relative splittings between the values of the energy density in the adjacent minima are of the order of $1/N$, while the probability of the false vacuum decay is proportional to $N^{-1} \exp(-N)$ [207, 208]. The n quanta (n quarks-solitons) interpolate between the adjacent minima.

The existence of a large family of quasi-vacua can be inferred from the study of the θ evolution of the theory. Consider the topological susceptibility, i.e. the correlation function of two topological densities

$$\int d^2x \, \langle Q(x), Q(0) \rangle, \tag{6.1.18}$$

where

$$Q = \frac{i}{2\pi} \varepsilon_{kp} \partial_k A_p = \frac{1}{2\pi} \varepsilon_{kp} \left(\partial_k n_\ell^* \, \partial_p n^\ell \right). \tag{6.1.19}$$

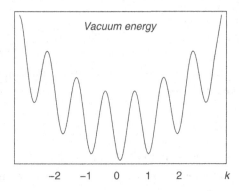

Figure 6.2. The vacuum structure of CP($N - 1$) model at $\theta = 0$.

The correlation function (6.1.18) is proportional to the second derivative of the vacuum energy with respect to the θ angle. From (6.1.19) it is not difficult to deduce that this correlation function scales as $1/N$ in the large N limit. The vacuum energy by itself scales as N. Thus, we conclude that, in fact, the vacuum energy should be a function of θ/N.

On the other hand, on general grounds, the vacuum energy must be a 2π-periodic function of θ. These two requirements are seemingly self-contradictory. A way out reconciling the above facts is as follows. Assume that we have a family of quasi-vacua with energies

$$E_k(\theta) \sim N \Lambda_\sigma^2 \left\{ 1 + \text{const} \left(\frac{2\pi k + \theta}{N} \right)^2 \right\}, \quad k = 0, \ldots, N - 1. \quad (6.1.20)$$

A schematic picture of these vacua is given in Fig. 6.2. All these minima are entangled in the θ evolution. If we vary θ continuously from 0 to 2π the depths of the minima "breathe." At $\theta = \pi$ two vacua become degenerate, while for larger values of θ the former global minimum becomes local while the adjacent local minimum becomes global. It is obvious that for the neighboring vacua which are not too far from the global minimum

$$E_{k+1} - E_k \sim \frac{\Lambda_\sigma^2}{N}. \quad (6.1.21)$$

This is also the confining force acting between n and $\bar{n}$.

One could introduce order parameters that would distinguish between distinct vacua from the vacuum family. An obvious choice is the expectation value of the topological charge. The kinks n^ℓ interpolate, say, between the global minimum and the first local one on the right-hand side. Then $\bar{n}$'s interpolate between the first local minimum and the global one. Note that the vacuum energy splitting is an effect suppressed by $1/N$. At the same time, these kinks have masses which scale as N^0,

$$M_\ell^{\text{kink}} \sim \Lambda_\sigma. \tag{6.1.22}$$

The multiplicity of such kinks is N [67], they form an N-plet of SU(N). This is in full accord with the fact that the large-N solution of (6.1.11) exhibits N quanta of the complex field n^ℓ.

Thus we see that the CP($N-1$) model has a fine structure of "vacua" which are split, with the splitting of the order of Λ_σ^2/N. In four-dimensional bulk theory these "vacua" correspond to elementary non-Abelian strings. Classically all these strings have the same tension (4.2.12). Due to quantum effects in the world sheet theory the degeneracy is lifted: the elementary strings become split, with the tensions

$$T = 2\pi\xi + c_1\, N\, \Lambda_\sigma^2 \left\{ 1 + c_2 \left(\frac{2\pi k + \theta}{N} \right)^2 \right\}, \tag{6.1.23}$$

where c_1 and c_2 are numerical coefficients. Note that (i) the splitting does not appear to any finite order in the coupling constant; (ii) since $\xi \gg \Lambda_\sigma$, the splitting is suppressed in both parameters, $\Lambda_\sigma/\sqrt{\xi}$ and $1/N$.

Kinks of the world-sheet theory represent confined monopoles (string junctions) in the four-dimensional bulk theory. Therefore kink confinement in CP($N-1$) model can be interpreted as follows [155, 154]. The non-Abelian monopoles, in addition to the four-dimensional confinement (which ensures that the monopoles are attached to the strings) acquire a two-dimensional confinement along the string: a monopole–antimonopole forms a meson-like configuration, with necessity, see Fig. 6.1.

In summary, the CP($N-1$) model in the Coulomb/confining phase, at small m, has a vacuum family with a fine structure. For each given θ (except $\theta = \pi, 3\pi$, etc.) the true ground state is unique, but there is a large number of "almost" degenerate ground states. The Z_N symmetry is unbroken. The classical condition (4.4.3) is replaced by $\langle n^\ell \rangle = 0$. The spectrum of physically observable states consists of kink-anti-kink mesons which form the adjoint representation of SU(N).

Instead, at large m the theory is in the Higgs phase; it has N strictly degenerate vacua (6.1.13); the Z_N symmetry is broken. We have $N-1$ elementary excitations n^ℓ with masses given by Eq. (6.1.14). Thus we conclude that these two regimes should be separated by a phase transition at some critical value m_* [154, 164]. This phase transition is associated with the Z_N symmetry breaking: in the Higgs phase the Z_N symmetry is spontaneously broken, while in the Coulomb phase it is restored. For $N = 2$ we deal with Z_2 which makes the situation akin to the Ising model.

In the world-sheet theory this is a phase transition between the Higgs and Coulomb/confining phase. In the bulk theory it can be interpreted as a phase

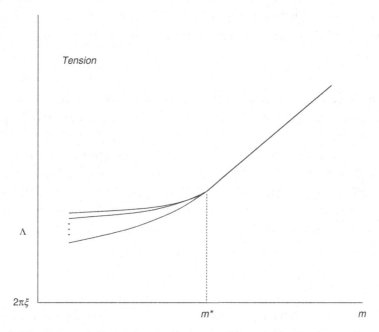

Figure 6.3. Schematic dependence of string tensions on the mass parameter m. At small m in the non-Abelian confinement phase the tensions are split while in the Abelian confinement phase at large m they are degenerative.

transition between the Abelian and non-Abelian confinement. In the Abelian confinement phase at large m, the Z_N symmetry is spontaneously broken, all N strings are strictly degenerate, and there is no two-dimensional confinement of the 4D-confined monopoles. In contrast, in the non-Abelian confinement phase occurring at small m, the Z_N symmetry is fully restored, all N elementary strings are split, and the 4D-confined monopoles combine with antimonopoles to form a meson-like configuration on the string, see Fig. 6.1. We show schematically the dependence of the string tensions on m in these two phases in Fig. 6.3.

In [164] the phase transition point is found using large-N methods developed by Witten in [159]. It turns out that the critical point is

$$m_* = \Lambda_\sigma. \tag{6.1.24}$$

The vacuum energy is calculated in both phases and is shown to be continuous at the critical point. If one approaches the critical point, say, from the Higgs phase some composite states of the world sheet theory (6.1.11), such as the photon and the kinks, become light. One is tempted to believe that these states become massless at the critical point (6.1.24). However, this happens only in a very narrow vicinity of the phase transition point where $1/N$ expansion fails. Thus, the large-N approximation is not powerful enough to determine the critical behavior.

To conclude this section we would like to stress that we encounter a crucial difference between the non-Abelian confinement in supersymmetric and non-supersymmetric gauge theories. For BPS strings in supersymmetric theories we have no phase transition separating the phase of the non-Abelian strings from that of the Abelian strings [132, 133]. Even for small values of the mass parameters supersymmetric theory strings are strictly degenerate, and the Z_N symmetry is spontaneously broken. In particular, at $\Delta m_A = 0$ the order parameter for the broken Z_N, which differentiates the N degenerate vacua of the supersymmetric CP($N-1$) model, is the bifermion condensate of two-dimensional fermions living on the string world sheet of the non-Abelian BPS string, see Section 4.4.3 and Section 5.1.3.

Moreover, the presence of the phase transition between Abelian and non-Abelian confinement in non-supersymmetric theories suggests a solution for the problem of enrichment of the hadronic spectrum mentioned in the beginning of Section 4, see also a more detailed discussion in Section 4.9. In the phase of Abelian confinement we have N strictly degenerative Abelian Z_N strings which give rise to too many "hadron" states, not present in actual QCD. Therefore, the Abelian Z_N strings can hardly play a role of prototypes for QCD confining strings. Although the BPS strings in supersymmetric theories become non-Abelian as we tune the mass parameters m_A to a common value, still there are N strictly degenerative non-Abelian strings and, therefore, still too many "hadron" states in the spectrum.

As was explained in this section, the situation in non-supersymmetric theories is quite different. As we make the mass parameters m_A equal we enter the non-Abelian confinement phase. In this phase N elementary non-Abelian strings are split. Say, at $\theta = 0$ we have only one lightest elementary string producing a single two-particle meson with the given flavor quantum numbers and spin, exactly as observed in nature. If N is large, the splitting is small, however. If N is not-so-large the splitting is of the order of Λ_σ^2. Therefore, the mesons produced by excited strings are unstable and may appear invisible experimentally.

6.2 Non-Abelian strings in $\mathcal{N} = 1^*$ theory

So far in our quest for the non-Abelian strings we focused on a particular model, with the $SU(N) \times U(1)$ gauge group and fundamental matter. However, it is known that solutions for Z_N strings were first found in simpler models, with the $SU(N)$ gauge group and adjoint matter [134, 135, 136, 137] (in fact, the gauge group becomes $SU(N)/Z_N$ if only adjoint matter is present in the theory). A natural question which immediately comes to one's mind is: can these Z_N strings under some special conditions develop orientational zero modes and become non-Abelian? The answer to this question is yes. Solutions for non-Abelian strings in the simplest theory with the $SU(2)$ gauge group and adjoint matter were found in [155] (actually, the gauge group of this theory is $SO(3)$). Here we will briefly review the results of this paper.

Although the model considered in [155] is supersymmetric the price one has to pay for its simplicity is that the strings which appear in this model are not BPS. The reason is easy to understand. One cannot introduce the FI term in the theory with the gauge group $SU(N)$ and, therefore, one cannot construct the string central charge [27].

The model considered in [155] is the so-called $\mathcal{N} = 1^*$ supersymmetric theory with the gauge group $SU(2)$. It is a deformed $\mathcal{N} = 4$ theory with the mass terms for three $\mathcal{N} = 1$ chiral superfields. Let us take two equal masses, say $m_1 = m_2 = m$, while the third mass m_3 is assumed to be distinct. Generally speaking, $\mathcal{N} = 4$ supersymmetry is broken down to $\mathcal{N} = 1$, unless $m_3 = 0$. If $m_3 = 0$ the theory has $\mathcal{N} = 2$ supersymmetry. It exemplifies $\mathcal{N} = 2$ gauge theory with the adjoint matter (two $\mathcal{N} = 1$ flavors of adjoint matter with equal masses).

Classically the vacuum structure of this theory was studied in [209], while quantum effects were taken into account in [210]. The $\mathcal{N} = 4$ theory with the $SU(2)$ gauge group has three vacua, and if the coupling constant of the $\mathcal{N} = 4$ theory is small,[3] $g^2 \ll 1$, one of these vacua is at weak coupling. All three adjoint scalars condense in this vacuum. Therefore, it is called the Higgs vacuum [209, 210]. Two other vacua of the theory are always at strong coupling. For small m_3 they correspond to the monopole and dyon vacua of the perturbed $\mathcal{N} = 2$ theory [2]. Here we will concentrate on the Higgs vacuum in the weak coupling regime.

In this vacuum the gauge group $SU(2)$ is broken down to Z_2 by the adjoint scalar VEV's. Therefore there are stable Z_2 non-BPS strings associated with

$$\pi_1(SU(2)/Z_2) = Z_2. \tag{6.2.1}$$

If we choose a special value of m_3,

$$m_3 = m,$$

---

[3] Note that the coupling of the unbroken $\mathcal{N} = 4$ theory g^2 does not run since the $\mathcal{N} = 4$ theory is conformal.

there is a diagonal $O(3)_{C+F}$ subgroup of the global gauge group SU(2), and the flavor O(3) group, unbroken by vacuum condensates. In parallel with Refs. [131, 130, 132, 133] (see also Chapter 4), the presence of this group leads to emergence of orientational zero modes of the Z_2-strings associated with rotation of the color magnetic flux of the string inside the SU(2) gauge group, which converts the Z_2 string into non-Abelian.

Let us discuss this model in more detail. In terms of $\mathcal{N} = 1$ supermultiplets, the $\mathcal{N} = 4$ supersymmetric gauge theory with the SU(2) gauge group contains a vector multiplet, consisting of the gauge field A_μ^a and gaugino $\lambda^{\alpha a}$, and three chiral multiplets Φ_A^a, $A = 1, 2, 3$, all in the adjoint representation of the gauge group, with $a = 1, 2, 3$ being the SU(2) color index. The superpotential of the $\mathcal{N} = 4$ gauge theory is

$$W_{\mathcal{N}=4} = -\frac{\sqrt{2}}{g^2} \varepsilon_{abc} \Phi_1^a \Phi_2^b \Phi_3^c. \tag{6.2.2}$$

One can deform this theory, breaking $\mathcal{N} = 4$ supersymmetry down to $\mathcal{N} = 2$, by adding two equal mass terms m, say, for the first two flavors of the adjoint matter,

$$W_{\mathcal{N}=2} = \frac{m}{2g^2} \sum_{A=1,2} (\Phi_A^a)^2. \tag{6.2.3}$$

Then, the third flavor combines with the vector multiplet to form an $\mathcal{N} = 2$ vector supermultiplet, while the first two flavors (6.2.3) can be treated as $\mathcal{N} = 2$ massive adjoint matter. If one wishes, one can further break supersymmetry down to $\mathcal{N} = 1$, by adding a mass term to the Φ_3 multiplet,

$$W_{\mathcal{N}=1^*} = \frac{m_3}{2g^2} (\Phi_3^a)^2. \tag{6.2.4}$$

The bosonic part of the action is

$$S_{\mathcal{N}=1^*} = \frac{1}{g^2} \int d^4x \left(\frac{1}{4} (F_{\mu\nu}^a)^2 + \sum_A |D_\mu \Phi_A^a|^2 \right.$$
$$+ \frac{1}{2} \sum_{A,B} [(\bar{\Phi}_A \bar{\Phi}_B)(\Phi_A \Phi_B) - (\bar{\Phi}_A \Phi_B)(\bar{\Phi}_B \Phi_A)]$$
$$\left. + \sum_A \left| \frac{1}{\sqrt{2}} \varepsilon_{abc} \varepsilon^{ABC} \Phi_B^b \Phi_C^c - m_A \Phi_A^a \right|^2 \right), \tag{6.2.5}$$

where $D_\mu \Phi_A^a = \partial_\mu \Phi_A^a + \varepsilon^{abc} A_\mu^b \Phi_A^c$, and we use the same notation Φ_A^a for the scalar components of the corresponding chiral superfields.

As was mentioned above, we are going to study the so-called Higgs vacuum of the theory (6.2.5), where all three adjoint scalars develop VEVs of the order of m, $\sqrt{mm_3}$. The scalar condensates Φ_A^a can be written in the form of the following 3×3 color-flavor matrix (convenient for the SU(2) gauge group and three chiral flavor superfields)

$$\langle\Phi_A^a\rangle = \frac{1}{\sqrt{2}}\begin{pmatrix} \sqrt{mm_3} & 0 & 0 \\ 0 & \sqrt{mm_3} & 0 \\ 0 & 0 & m \end{pmatrix}. \tag{6.2.6}$$

These VEV's break the SU(2) gauge group completely. The W-bosons masses are

$$m_{1,2}^2 = m^2 + mm_3 \tag{6.2.7}$$

for $A_\mu^{1,2}$, while the mass of the photon field A_μ^3 is

$$m_\gamma^2 = 2mm_3. \tag{6.2.8}$$

In what follows, we will be especially interested in a particular point in the parameter space: $m_3 = m$. For this value of m_3, (6.2.6) presents a symmetric color-flavor locked vacuum

$$\langle\Phi_A^a\rangle = \frac{m}{\sqrt{2}}\begin{pmatrix} 1 & 0 & 0 \\ 0 & 1 & 0 \\ 0 & 0 & 1 \end{pmatrix}. \tag{6.2.9}$$

This symmetric vacuum respects the global O(3)$_{C+F}$ symmetry,

$$\Phi \to O\Phi O^{-1}, \quad A_\mu^a \to O^{ab}A_\mu^b, \tag{6.2.10}$$

which combines transformations from the global color and flavor groups, similarly to the SU(N)$_{C+F}$ group of the U(N) theories, see Chapter 4. It is this symmetry that is responsible for the presence of the non-Abelian strings in the vacuum (6.2.9).

Note that at $m_3 = m$ all gauge bosons have equal masses,

$$m_g^2 = 2m^2, \tag{6.2.11}$$

as is clearly seen from (6.2.7) and (6.2.8). This means, in particular, that in the point $m_3 = m$ we lose all traces of the "Abelization" in our theory, which are otherwise present at generic values of m_3.

Let us also emphasize that the coupling g^2 in Eq. (6.2.5) is the $\mathcal{N} = 4$ coupling constant. It does not run in the $\mathcal{N} = 4$ theory at scales above m, and we assume it to be small,

$$g^2 \ll 1. \tag{6.2.12}$$

At the scale m the gauge group SU(2) is broken in the vacuum (6.2.9) by the scalar VEVs. Much in the same way as in the U(N) theory (see Chapter 4), the running of the coupling constant below the scale m is determined by the β function of the effective two-dimensional sigma model on the world sheet of the non-Abelian string.

Skipping details we present here the solution for the non-Abelian string in the model (6.2.5) found in [155]. When m_3 approaches m, the theory acquires additional symmetry. In this case the scalar VEVs take the form (6.2.9), preserving the global combined color-flavor symmetry (4.1.15). On the other hand, the Z_2 string solution itself is not invariant under this symmetry. The symmetry (4.1.15) generates orientational zero modes of the string. The string solution in the singular gauge is

$$\Phi_A^a = O \begin{pmatrix} \frac{g}{\sqrt{2}}\phi & 0 & 0 \\ 0 & \frac{g}{\sqrt{2}}\phi & 0 \\ 0 & 0 & a_0 \end{pmatrix} O^{-1}$$

$$= \frac{g}{\sqrt{2}}\phi\delta_A^a + S^a S^A \left(a_0 - \frac{g}{\sqrt{2}}\phi \right),$$

$$A_i^a = S^a \frac{\varepsilon_{ij}x_j}{r^2} f(r), \quad i, j = 1, 2, \tag{6.2.13}$$

where we introduced the unit orientational vector S^a,

$$S^a = O_b^a \delta^{b3} = O_3^a. \tag{6.2.14}$$

It is easy to see that the orientational vector S^a defined above coincides with the one we introduced in Section 4, see Eq. (4.4.21). The solution (6.2.13) interpolates between the Abelian Z_2 strings for which $\vec{S} = \{0, 0, \pm 1\}$. We see that the string flux is determined now by an arbitrary vector S^a in the color space, much in the same way as for the non-Abelian strings in the U(N) theories.

Since this string is not BPS-saturated, the profile functions in (6.2.13) satisfy now the *second*-order differential equations,

$$\phi'' + \frac{1}{r}\phi' - \frac{1}{r^2}f^2\phi = \phi\left(g^2\phi^2 - \sqrt{2}m_3 a_0\right) + 2\phi\left(a_0 - \frac{m}{\sqrt{2}}\right)^2,$$

$$a_0'' + \frac{1}{r}a_0' = -\frac{m_3}{\sqrt{2}}\left(g^2\phi^2 - \sqrt{2}m_3 a_0\right) + 2g^2\phi^2\left(a_0 - \frac{m}{\sqrt{2}}\right),$$

$$f'' - \frac{1}{r}f' = 2g^2 f\phi^2, \qquad\qquad\qquad\qquad\qquad (6.2.15)$$

where the primes stand for derivatives with respect to r, and the boundary conditions are

$$\phi(0) = 0, \quad \phi(\infty) = \frac{\sqrt{mm_3}}{g},$$

$$a_0'(0) = 0, \quad a(\infty) = \frac{m}{\sqrt{2}},$$

$$f(0) = 1, \quad f(\infty) = 0. \qquad\qquad\qquad (6.2.16)$$

The string tension is

$$T = 2\pi \int_0^\infty r\,dr\left[\frac{f'^2}{2g^2 r^2} + \phi'^2 + \frac{a_0'^2}{g^2} + \frac{f^2\phi^2}{r^2}\right.$$

$$\left. + \frac{g^2}{2}\left(\phi^2 - \frac{\sqrt{2}m_3}{g^2}a_0\right)^2 + 2\phi^2\left(a_0 - \frac{m}{\sqrt{2}}\right)^2\right].$$

$$(6.2.17)$$

The second-order equations for the string profile functions were solved in [155] numerically and the string tension was found as a function of the mass ratio m_3/m. Note that for the BPS string (which appears in the limit $m_3 \to 0$) the tension is

$$T_{\text{BPS}} = 2\pi \, mm_3/g^2.$$

The effective world sheet theory for the non-Abelian string (6.2.13) was shown to be the non-supersymmetric CP(1) model [155]. Its coupling constant β is related to the coupling constant g^2 of the bulk theory via (4.4.10), where now the normalization integral

$$I \sim 0.78.$$

In this theory there is a 't Hooft–Polyakov monopole with the unit magnetic charge. Since the Z_2-string charge is $1/2$, it cannot end on the monopole, much in the same way as for the monopoles in the U(N) theories, see Section 4.5. Instead, the confined monopole appears to be a junction of the Z_2 string and anti-string. In the world sheet CP(1) model it is seen as a kink interpolating between the two vacua.

At small values of the mass difference $m_3 - m$ the world sheet theory is in the Coulomb/confining phase, see Section 6.1.2, although, strictly speaking, the large-N analysis is not applicable in this case. Still, the monopoles, in addition to four-dimensional confinement ensuring that they are attached to a string, also experience confinement in two dimensions, along the string [155]. This means that each monopole on the string must be accompanied by an antimonopole, with a linear potential between them along the string. As a result, they form a meson-like configuration, see Fig. 6.1. As was mentioned in Section 6.1.2, this follows from the exact solution of the CP(1) model [204, 205]: only the triplets of SU(2)$_{C+F}$ are seen in the spectrum.

7

Strings on the Higgs branches

One common feature of supersymmetric gauge theories is the presence of moduli spaces – manifolds on which scalar fields can develop arbitrary VEVs without violating the zero energy condition. If on these vacuum manifolds the gauge group is broken, either completely or partially, down to a discrete subgroup, these manifolds are referred to as the Higgs branches.

One may pose a question: what happens with the flux tubes and confinement in theories with the Higgs branches? The Higgs branch represents an extreme case of type-I superconductivity, with vanishing Higgs mass. One may ask oneself whether or not the ANO strings still exist in this case, and if yes, whether they provide confinement for external heavy sources.

This question was posed and studied first in [102] where the authors concluded that the vortices do not exist on the Higgs branches due to infrared problems. In Refs. [211, 212] the $\mathcal{N} = 1$ SQED vortices were further analyzed. It was found that at a generic point on the Higgs branch strings are unstable. The only vacuum which supports string solutions is the base point of the Higgs branch where the strings become BPS-saturated. The so-called "vacuum selection rule" was put forward in [211, 212] to ensure this property.

On the other hand, in [103, 175] it was shown that infrared problems can be avoided provided certain infrared regularizations are applied. Say, in [103, 175] the infrared divergences were regularized through embedding of $\mathcal{N} = 1$ SQED in softly broken $\mathcal{N} = 2$ SQED. Another alternative is to consider a finite length-L string instead of an infinitely long string. In this case the impact of the Higgs branch was shown to "roughen" the string, making it logarithmically "thick." Still, the string solutions do exist and produce confinement for heavy trial sources. However, now the confining potential is not linear in separation; rather it behaves as

$$V(L) \sim \frac{L}{\ln L}$$

at large L. Below we will briefly review the string solutions on the Higgs branches, starting from the simplest case of the flat Higgs branch and then considering a more common scenario, when the Higgs branch is curved by the FI term.

7.1 Extreme type-I strings

In this section we will review the classical solutions for the ANO vortices (flux tubes) in the theories with the flat Higgs potential which arises in supersymmetric settings [103]. Let us start from the Abelian Higgs model,

$$S_{\text{AH}} = \int d^4x \left\{ \frac{1}{4g^2} F_{\mu\nu}^2 + |\nabla_\mu q|^2 + \lambda \left(|q|^2 - v^2 \right)^2 \right\}, \qquad (7.1.1)$$

for a single complex field q with the quartic coupling $\lambda \to 0$. Here

$$\nabla_\mu = \partial_\mu - i n_e A_\mu,$$

where n_e is the electric charge of the field q. Following [103], we will first consider this model with a small but nonvanishing λ and then take the limit $\lambda = 0$.

Obviously, the field q develops a VEV, $q = v$, spontaneously breaking the U(1) gauge group. The photon acquires the mass

$$m_\gamma^2 = 2n_e^2 g^2 v^2, \qquad (7.1.2)$$

while the Higgs particle mass is

$$m_q^2 = 4\lambda v^2. \qquad (7.1.3)$$

The model (7.1.1) is the standard Abelian Higgs model which supports the ANO strings [36]. For generic values of λ the Higgs mass differs from that of the photon. The ratio of the photon mass to the Higgs mass is an important parameter – in the theory of superconductivity it characterizes the type of the superconductor in question. Namely, for $m_q < m_\gamma$ we have the type-I superconductor in which two well-separated ANO strings attract each other. On the other hand, for $m_q > m_\gamma$ we have the type-II superconductor in which two well-separated strings repel each other. This is due to the fact that the scalar field gives rise to attraction between two vortices, while the electromagnetic field gives rise to repulsion.

Now, let us consider the extreme type-I limit in which

$$m_q \ll m_\gamma. \qquad (7.1.4)$$

We will assume the weak coupling regime in the model (7.1.1), $\lambda \ll g^2 \ll 1$.

The general guiding idea which will lead us in the search for the string solution in the extreme type-I limit is a separation of different fields at distinct scales which are obviously present in the problem at hand due to the "extremality" condition (7.1.4). This method goes back to the original paper by Abrikosov [36] in which the tension of the type-II string had been calculated under the condition $m_q \gg m_\gamma$. A similar idea was used in [103] to calculate the tension of the type-I string under the condition $m_q \ll m_\gamma$.

To the leading order in $\ln m_\gamma / m_q$ the vortex solution has the following structure in the plane orthogonal to the string axis: The electromagnetic field is confined in a core with the radius

$$R_g \sim \frac{1}{m_\gamma} \ln \frac{m_\gamma}{m_q}. \qquad (7.1.5)$$

At the same time, the scalar field is close to zero inside the core. Outside the core the electromagnetic field is vanishingly small, while the scalar field behaves as

$$q = v \left\{ 1 - \frac{K_0(m_q r)}{\ln(1/m_q R_g)} \right\} e^{i\alpha}, \qquad (7.1.6)$$

where r and α are polar coordinates in the orthogonal plane (Fig. 3.6). Here K_0 is the (imaginary argument) Bessel function[1] with the exponential fall-off at infinity and logarithmic behavior at small arguments,

$$K_0(x) \sim \ln(1/x) \text{ at } x \to 0.$$

The reason for this behavior is that in the absence of the electromagnetic field outside the core the scalar field satisfies the free equation of motion, and (7.1.6) presents the appropriate solution to this equation. From (7.1.6) we see that the scalar field slowly (logarithmically) approaches its boundary value v.

The tension of this string is [103]

$$T = \frac{2\pi v^2}{\ln (m_\gamma / m_q)}. \qquad (7.1.7)$$

The main contribution to the tension in (7.1.7) comes from the logarithmic "tail" of the scalar field q. It is given by the kinetic term for the scalar field in (7.1.1). This term contains a logarithmic integral over r. Other terms in the action are

---

[1] It is also known as the McDonald function.

suppressed by inverse powers of $\ln(m_\gamma/m_q)$ as compared with the contribution quoted in (7.1.7).

The results in Eqs. (7.1.5) and (7.1.7) imply that if we naively take the limit $m_q \to 0$ the string becomes infinitely thick and its tension tends to zero [103]. This apparently means that there are no strings in the limit $m_q = 0$. As was mentioned above, the absence of the ANO strings in the theories with the flat Higgs potential was first noted in [102].

One might think that the absence of the ANO strings means that there is no confinement of monopoles in the theories with the Higgs branches.

We hasten to say that this is a wrong conclusion.

As we will see shortly confinement does not disappear [103]. It is the formulation of the problem that has to be changed a little bit in the case at hand.

So far we considered infinitely long ANO strings. However, an appropriate setup in the confinement problem is in fact slightly different [103]. We have to consider a monopole–antimonopole pair at a large but finite separation L. Our aim is to take the limit $m_q \to 0$. This limit will be perfectly smooth provided we consider the ANO string of a finite length L, such that

$$\frac{1}{m_\gamma} \ll L \ll \frac{1}{m_q}. \tag{7.1.8}$$

Then it turns out [103] that $1/L$ plays the role of the infrared (IR) cutoff in Eqs. (7.1.5) and (7.1.7), rather than m_q. The reason for this is that for $r \ll L$ the problem is two-dimensional and the solution of the two-dimensional free equation of motion for the scalar field given by (7.1.6) is logarithmic. If we naively put $m_q = 0$ in this solution the McDonald function reduces to the logarithmic function which cannot reach a finite boundary value at infinity. Thus, as we mentioned above, infinitely long flux tubes do not exist.

However, for $r \gg L$, the problem becomes three-dimensional. The solution to the three-dimensional free scalar equation of motion behaves as

$$(q - v) \sim 1/|\vec{x}|$$

where x_n ($n = 1, 2, 3$) are the spatial coordinates in the three-dimensional space.

With this behavior the scalar field reaches its boundary value at infinity. Clearly, $1/L$ plays the role of an IR cutoff for the logarithmic behavior of the scalar field.

Now we can safely put $m_q = 0$. The formula for the radius of the electromagnetic core of the vortex takes the form

$$R_g \sim \frac{1}{m_\gamma} \ln(m_\gamma L), \tag{7.1.9}$$

while the string tension now becomes [103]

$$T = \frac{2\pi v^2}{\ln\left(m_\gamma L\right)}.$$
(7.1.10)

The ANO string becomes "thick." Nevertheless, its transverse size R_g is much smaller than its length L,

$$R_g \ll L,$$

so that the string-like structure is clearly identifiable. As a result, the potential acting between the probe well-separated monopole and antimonopole confines but is no longer linear in L. At large L [103]

$$V(L) = 2\pi v^2 \frac{L}{\ln\left(m_\gamma L\right)}.$$
(7.1.11)

The potential $V(L)$ is an order parameter which distinguishes different phases of a given gauge theory (see, for example, [71]). We conclude that on the Higgs branches one deals with a new confining phase, which had never been observed previously. It is clear that this phase can arise only in supersymmetric theories because we have no Higgs branches without supersymmetry.

7.2 Example: $\mathcal{N} = 1$ SQED with the FI term

Initial comments regarding this model are presented in Part I, see Section 3.2.2. The SQED Lagrangian in terms of superfields is presented in Eq. (3.2.1), while the component expression can be found in (3.2.5). For convenience we reiterate here crucial features of $\mathcal{N} = 1$ SQED, to be exploited below.

The field content of $\mathcal{N} = 1$ SQED is as follows. The vector multiplet contains the U(1) gauge field A_μ and the Weyl fermion λ^α, $\alpha = 1, 2$. The chiral matter multiplet contains two complex scalar fields q and $\tilde{q}$ as well as two complex Weyl fermions ψ^α and $\tilde{\psi}_\alpha$. The bosonic part of the action is

$$S_{\text{SQED}} = \int d^4x \left\{ \frac{1}{4g^2} F_{\mu\nu}^2 + \bar{\nabla}_\mu \bar{q} \nabla_\mu q + \bar{\nabla}_\mu \tilde{q} \nabla_\mu \bar{\tilde{q}} + V(q, \tilde{q}) \right\}, \quad (7.2.1)$$

where

$$\nabla_\mu = \partial_\mu - \frac{i}{2} A_\mu, \quad \bar{\nabla}_\mu = \partial_\mu + \frac{i}{2} A_\mu.$$

Thus, we assume the matter fields to have electric charges $n_e = \pm 1/2$. The scalar potential of this theory comes from the D term and reduces to

$$V(q, \tilde{q}) = \frac{g^2}{8} (|q|^2 - |\tilde{q}|^2 - \xi)^2. \quad (7.2.2)$$

The parameter ξ is the Fayet–Iliopoulos parameter introduced through §3.

The vacuum manifold of the theory (7.2.1) is the Higgs branch determined by the condition

$$|q|^2 - |\tilde{q}|^2 = \xi. \quad (7.2.3)$$

The dimension of this Higgs branch is two. To see this please observe that in the problem at hand we have two complex scalars (four real variables) subject to one constraint (7.2.3). In addition, we have to subtract one gauge phase; thus, we have $4 - 1 - 1 = 2$.

In general, the physics of the massless modes in theories with the Higgs branches can be described in terms of an effective low-energy sigma model

$$S_{\text{LE}} = \int d^4x \, g_{MN}(\varphi) \partial_\mu \varphi^N \partial_\mu \varphi^M, \quad (7.2.4)$$

where φ^M are massless scalar fields parametrizing the given Higgs branch and g_{MN} is the metric which depends on φ.

For example, the squark fields in $\mathcal{N} = 1$ SQED subject to the constraint (7.2.3) can be parametrized as follows:

$$\begin{aligned} q &= \sqrt{\xi} \, e^{i\alpha + i\beta} \cosh \rho, \\ \bar{\tilde{q}} &= \sqrt{\xi} \, e^{i\alpha - i\beta} \sinh \rho, \end{aligned} \quad (7.2.5)$$

where α is an (irrelevant) gauge phase while $\rho(x)$ and $\beta(x)$ are two massless fields living on the Higgs branch. With this parametrization the sigma model (7.2.4) on the Higgs branch takes the form [175]

$$S_{\text{LE}} = \xi \int d^4x \left\{ \cosh 2\rho \left[(\partial_\mu \rho)^2 + (\partial_\mu \beta)^2 \tanh^2 2\rho \right] \right\}. \quad (7.2.6)$$

From this expression one can immediately read off the two-by-two metric tensor.

The mass spectrum of $\mathcal{N} = 1$ SQED with the FI term, as it is defined in Eqs. (7.2.1) and (7.2.2), consists of one massive vector $\mathcal{N} = 1$ multiplet, with mass

$$m_\gamma^2 = \frac{1}{2} g^2 v^2, \tag{7.2.7}$$

(four bosonic + four fermionic states) and one chiral massless field associated with fluctuations along the Higgs branch. The VEV of the scalar field above is given by

$$v^2 = |\langle q \rangle|^2 + |\langle \tilde{q} \rangle|^2. \tag{7.2.8}$$

Next, following [175], let us consider strings supported by this theory. First we will choose the scalar field VEV to lie on the base point of the Higgs branch,

$$q = \sqrt{\xi}, \quad \tilde{q} = 0. \tag{7.2.9}$$

Then the massless field $\tilde{q}$ plays no role in the string solution and can be set to zero. This case is similar to the case of non-Abelian strings in $\mathcal{N} = 1$ SQCD described in detail in Section 5.1. On the base of the Higgs branch we do have (classically) the BPS ANO strings with the tension given by (4.2.12). In particular, their profile functions are determined by (3.2.18) and satisfy the first-order equations (3.2.19).

Now consider a generic vacuum on the Higgs branch. The string solution has the following structure [175]. The electromagnetic field, together with the massive scalar, form a string core of size $\sim 1/(g\sqrt{\xi})$. The solution for this core is essentially given by the BPS profile functions for the gauge field and massive scalar q. Outside the core the massive fields almost vanish, while the light (massless) fields living on the Higgs branch produce a logarithmic "tail." Inside this "tail" the light scalar fields interpolate between the base point (7.2.9) and the VEVs of scalars q and $\tilde{q}$ on the Higgs branch (7.2.3). The tension of the string is given by the sum of tensions coming from the core and "tail" regions,

$$T = 2\pi\xi + \frac{2\pi\xi}{\ln(g\sqrt{\xi}\, L)} l^2, \tag{7.2.10}$$

where l is the length of the geodesic line on the Higgs branch between the base point and the VEV,

$$l = \int_0^1 dt \sqrt{g_{MN} \left(\partial_t \varphi^N\right)\left(\partial_t \varphi^N\right)}, \tag{7.2.11}$$

where g_{MN} is the metric on the Higgs branch, while φ^N stand for massless scalars living on the Higgs branch (see e.g. (7.2.6)). For example, for $v^2 \gg \xi$

$$l^2 = v^2/\xi,$$

and the "tail" contribution in (7.2.10) matches the result (7.1.10) for the string tension on the flat Higgs branch.

In (7.2.10) we consider the string of a finite length L to ensure infrared regularization. It is also possible [175] to embed $\mathcal{N} = 1$ SQED (7.2.1) in softly broken $\mathcal{N} = 2$ SQED much in the same way as it was done in Section 5.1 for non-Abelian strings. This procedure slightly lifts the Higgs branch making even infinitely long strings well defined. Note, however, that within this procedure the string is *not* BPS-saturated at a generic point on the Higgs branch.

8

Domain walls as *D*-brane prototypes

D branes are extended objects in string theory on which strings can end [10]. Moreover, the gauge fields are the lowest excitations of open superstrings, with the endpoints attached to *D* branes. SU(N) gauge theories are obtained as a field-theoretic reduction of a string theory on the world volume of a stack of N *D* branes.

Our task is to see how the above assertions are implemented in field theory. We have already thoroughly discussed field-theoretic strings. Solitonic objects of the domain wall type were also extensively studied in supersymmetric gauge theories in 1+3 dimensions. The original impetus was provided by the Dvali–Shifman observation [11] of the critical (BPS-saturated) domain walls in $\mathcal{N} = 1$ gluodynamics, with the tension scaling as $N\Lambda^3$. The peculiar N dependence of the tension prompted [12] a *D*-brane interpretation of such walls. Ideas as to how flux tubes can end on the BPS walls were analyzed [213] at the qualitative level shortly thereafter. Later on, BPS-saturated domain walls and their junctions with strings were discussed [214, 215] in a more quantitative aspect in $\mathcal{N} = 2$ sigma models. Some remarkable parallels between field-theoretical critical solitons and the *D*-brane string theory construction were discovered.

In this and subsequent chapters we will review the parallel found between the field-theoretical BPS domain walls in gauge theories and *D* branes/strings. In other words, we will discuss BPS domain walls with the emphasis on localization of the gauge fields on their world volume. In this sense the BPS domain walls become *D*-brane prototypes in field theory.

As was mentioned, research on field-theoretic mechanisms of gauge field localization on the domain walls attracted much attention. The only viable mechanism of gauge field localization was outlined in Ref. [11] where it was noted that if a gauge field is confined in the bulk and is unconfined (or less confined) on the brane, this naturally gives rise to a gauge field on the wall (for further developments see Refs. [216, 217]). Although this idea seems easy to implement, in fact it requires a careful consideration of quantum effects (confinement is certainly such an effect) which is hard to do at strong coupling.

Building on these initial proposals models with localization of gauge fields on the world volume of domain walls at weak coupling in $\mathcal{N} = 2$ supersymmetric gauge theories were suggested in [142, 37, 218]. Using a dual language, the basic idea can be expressed as follows: the gauge group is completely Higgsed in the bulk while inside the wall the charged scalar fields almost vanish. In the bulk magnetic flux tubes are formed while inside the wall the magnetic fields can propagate freely. In Ref. [142] domain walls in the simplest $\mathcal{N} = 2$ SQED theory were considered while Refs. [218, 37, 219] deal with the domain walls in non-Abelian $\mathcal{N} = 2$ gauge theories (4.1.7), with the gauge group U(N). Below we will review some results obtained in these papers.

The moduli space of the multiple domain walls in $\mathcal{N} = 2$ supersymmetric gauge theories and corresponding sigma models were studied in [220, 221, 222, 223, 224]. Note that the domain walls can intersect [84, 85, 88]. In particular, in [86, 87] honeycomb webs of walls were obtained in Abelian and non-Abelian gauge theories, respectively. We briefly discussed this phenomenon in Part I, Section 3.1.5.

We start our discussion of the BPS domain walls as D-brane prototypes in the simplest Abelian theory – $\mathcal{N} = 2$ SQED with 2 flavors [142]. It supports both the BPS-saturated domain walls and the BPS-saturated ANO strings if the Fayet–Iliopoulos term is added to the theory.

8.1 $\mathcal{N} = 2$ supersymmetric QED

$\mathcal{N} = 1$ SQED (four supercharges) was discussed in Section 3.2. Now we will extend supersymmetry to $\mathcal{N} = 2$ (eight supercharges). Some relevant features of this model are summarized in Appendix C.

The field content of $\mathcal{N} = 2$ SQED is as follows. In the gauge sector we have the U(1) vector $\mathcal{N} = 2$ multiplet. In the matter sector we have N_f matter hypermultiplets. In this section we will limit ourselves to $N_f = 2$. This is the simplest case which admits domain wall interpolating between quark vacua. The bosonic part of the action of this theory is

$$
S = \int d^4x \left\{ \frac{1}{4g^2} F^2_{\mu\nu} + \frac{1}{g^2} |\partial_\mu a|^2 + \bar{\nabla}_\mu \bar{q}_A \nabla_\mu q^A + \bar{\nabla}_\mu \tilde{q}_A \nabla_\mu \bar{\tilde{q}}^A \right.
$$
$$
\left. + \frac{g^2}{8} (|q^A|^2 - |\tilde{q}_A|^2 - \xi)^2 + \frac{g^2}{2} |\tilde{q}_A q^A|^2 + \frac{1}{2} (|q^A|^2 + |\tilde{q}^A|^2)|a + \sqrt{2} m_A|^2 \right\},
$$
$$
(8.1.1)
$$

where

$$\nabla_\mu = \partial_\mu - \frac{i}{2} A_\mu, \quad \bar{\nabla}_\mu = \partial_\mu + \frac{i}{2} A_\mu. \tag{8.1.2}$$

With this convention the electric charges of the matter fields are $\pm 1/2$ (in the units of g). Parameter ξ in Eq. (8.1.1) is the coefficient in front of the Fayet–Iliopoulos term. It is introduced as in Eq. (4.1.5) with $F_3 = D$ and $F_{1,2} = 0$. In other words, here we introduce the Fayet–Iliopoulos term as the D term. Furthermore, g is the U(1) gauge coupling. The index $A = 1, 2$ is the flavor index.

The mass parameters m_1, m_2 are assumed to be real. In addition we will assume

$$\Delta m \equiv m_1 - m_2 \gg g\sqrt{\xi}. \tag{8.1.3}$$

Simultaneously, $\Delta m \ll (m_1 + m_2)/2$. There are two vacua in this theory: in the first vacuum

$$a = -\sqrt{2}m_1, \quad q_1 = \sqrt{\xi}, \quad q_2 = 0, \tag{8.1.4}$$

and in the second one

$$a = -\sqrt{2}m_2, \quad q_1 = 0, \quad q_2 = \sqrt{\xi}. \tag{8.1.5}$$

The vacuum expectation value of the field $\tilde{q}$ vanishes in both vacua. Hereafter in the search for domain wall solutions we will stick to the *ansatz* $\tilde{q} = 0$.

Now let us discuss the mass spectrum in both quark vacua. Consider for definiteness the first vacuum, Eq. (8.1.4). The spectrum can be obtained by diagonalizing the quadratic form in (8.1.1). This is done in Ref. [35]; the result is as follows: one real component of the field q^1 is eaten up by the Higgs mechanism to become the third component of the massive photon. Three components of the massive photon, one remaining component of q^1 and four real components of the fields $\tilde{q}_1$ and a form one long $\mathcal{N} = 2$ multiplet (8 boson states $+$ 8 fermion states), with mass

$$m_\gamma^2 = \frac{1}{2} g^2 \xi. \tag{8.1.6}$$

The second flavor q^2, $\tilde{q}_2$ (which does not condense in this vacuum) forms one short $\mathcal{N} = 2$ multiplet (4 boson states $+$ 4 fermion states), with mass Δm which is heavier than the mass of the vector supermultiplet. The latter assertion applies to the regime (8.1.3). In the second vacuum the mass spectrum is similar – the roles of the first and the second flavors are interchanged.

If we consider the limit opposite to that in Eq. (8.1.3) and tend $\Delta m \to 0$, the "photonic" supermultiplet becomes heavier than that of q^2, the second flavor field.

Therefore, it can be integrated out, leaving us with the theory of massless moduli from q^2, $\tilde{q}_2$, which interact through a nonlinear sigma model with the Kähler term corresponding to the Eguchi–Hanson metric. The manifold parametrized by these (nearly) massless fields is obviously four-dimensional. Both vacua discussed above lie at the base of this manifold. Therefore, in considering the domain wall solutions in the sigma model limit $\Delta m \to 0$ [220, 221, 215] one can limit oneself to the base manifold, which is, in fact, a two-dimensional sphere. In other words, classically, it is sufficient to consider the domain wall in the CP(1) model deformed by a twisted mass term (related to a nonvanishing Δm), see Fig. 3.11. This was first done in [221]. A more general analysis of the domain walls on the Eguchi–Hanson manifold can be found in [225]. An interesting $\mathcal{N} = 1$ deformation of the model (8.1.1) which was treated in the literature [226] in the quest for "confinement on the wall" automatically requires construction of the wall on the Eguchi–Hanson manifold, rather than the CP(1) wall, since in this case the two vacua of the model between which the wall interpolates do not lie on the base.

8.2 Domain walls in $\mathcal{N} = 2$ SQED

A BPS domain wall interpolating between the two vacua of the bulk theory (8.1.1) was explicitly constructed in Ref. [142]. Assuming that all fields depend only on the coordinate $z = x_3$, it is possible to write the energy by performing the Bogomol'nyi completion [5],

$$
E = \int dx_3 \left\{ \left| \nabla_3 q^A \pm \frac{1}{\sqrt{2}} q^A (a + \sqrt{2} m_A) \right|^2 \right.
$$
$$
\left. + \left| \frac{1}{g} \partial_3 a \pm \frac{g}{2\sqrt{2}} \left(|q^A|^2 - \xi \right) \right|^2 \pm \frac{1}{\sqrt{2}} \xi \partial_3 a \right\}. \qquad (8.2.1)
$$

Requiring the first two terms above to vanish gives us the BPS equations for the wall. Assuming that $\Delta m > 0$ we choose the upper sign in (8.2.1) to get

$$
\nabla_z q^A = -\frac{1}{\sqrt{2}} q^A \left(a + \sqrt{2} m_A \right),
$$
$$
\partial_z a = -\frac{g^2}{2\sqrt{2}} \left(|q^A|^2 - \xi \right). \qquad (8.2.2)
$$

These first-order equations should be supplemented by the following boundary conditions:

$$
q^1(-\infty) = \sqrt{\xi}, \quad q^2(-\infty) = 0, \quad a(-\infty) = -\sqrt{2} m_1;
$$
$$
q^1(\infty) = 0, \quad |q^2(\infty)| = \sqrt{\xi}, \quad a(\infty) = -\sqrt{2} m_2, \qquad (8.2.3)
$$

which show that our wall interpolates between the two quark vacua. Here we use a U(1) gauge rotation to make q^1 in the left vacuum real.

The tension is given by the total derivative term (the last one in Eq. (8.2.1)) which can be identified as the $(1,0)$ central charge of the supersymmetry algebra,

$$T_{\mathrm{w}} = \xi \, \Delta m. \tag{8.2.4}$$

We can find the solution to the first-order equations (8.2.2) compatible with the boundary conditions (8.1.3). The range of variation of the field a inside the wall is of the order of Δm (see Eq. (8.2.3)). Minimization of its kinetic energy implies that this field slowly varies. Therefore, we may safely assume that the wall is thick; its size $R \gg 1/g\sqrt{\xi}$. This fact will be confirmed shortly.

We arrive at the following picture of the domain wall at hand. The wall solution has a three-layer structure [142], see Fig. 8.1. In the two outer layers – let us call them edges, they have thickness $O((g\sqrt{\xi})^{-1})$ which means that they are thin – the squark fields drop to zero exponentially; in the inner layer the field a interpolates between its two vacuum values.

Then to the leading order we can put the quark fields to zero in (8.2.2) inside the inner layer. The second equation in (8.2.2) tells us that a is a linear function of z. The solution for a takes the form

$$a = -\sqrt{2}\left(m - \Delta m \frac{z - z_0}{R}\right), \tag{8.2.5}$$

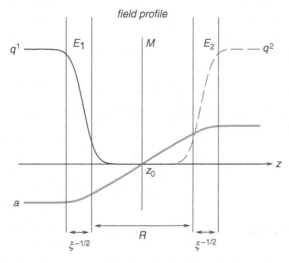

Figure 8.1. Internal structure of the domain wall: two edges (domains $E_{1,2}$) of the width $\sim (g\sqrt{\xi})^{-1}$ are separated by a broad middle band (domain M) of the width R, see Eq. (8.2.7).

where the collective coordinate z_0 is the position of the wall center (and Δm is assumed positive). The solution is valid in a wide domain of z

$$|z - z_0| < \frac{R}{2}, \tag{8.2.6}$$

except narrow areas of size $\sim 1/g\sqrt{\xi}$ near the edges of the wall at $z - z_0 = \pm R/2$.

Substituting the solution (8.2.5) in the second equation in (8.2.2) we get

$$R = \frac{4\Delta m}{g^2 \xi} = \frac{2\Delta m}{m_\gamma^2}. \tag{8.2.7}$$

Since $\Delta m/g\sqrt{\xi} \gg 1$, see Eq. (8.1.3), this result shows that $R \gg 1/g\sqrt{\xi}$, which justifies our approximation. This approximation will be referred to as the thin-edge approximation.

Furthermore, we can now use the first relation in Eq. (8.2.2) to determine tails of the quark fields inside the wall. As was mentioned above, we fix the gauge imposing the condition that q^1 is real at $z \to -\infty$, see a more detailed discussion in [142].

Consider first the left edge (domain E_1 in Fig. 8.1) at $z - z_0 = -R/2$. Substituting the above solution for a in the equation for q^1 we get

$$q^1 = \sqrt{\xi}\, e^{-\frac{m_\gamma^2}{4}\left(z - z_0 + \frac{R}{2}\right)^2}, \tag{8.2.8}$$

where m_γ is given by (8.1.6). This behavior is valid in the domain M, at $(z - z_0 + R/2) \gg 1/g\sqrt{\xi}$, and shows that the field of the first quark flavor tends to zero exponentially inside the wall, as was expected.

By the same token, we can consider the behavior of the second quark flavor near the right edge of the wall at $z - z_0 = R/2$. The first equation in (8.2.2) for $A = 2$ implies

$$q^2 = \sqrt{\xi}\, e^{-\frac{m_\gamma^2}{4}\left(z - z_0 - \frac{R}{2}\right)^2 - i\sigma}, \tag{8.2.9}$$

which is valid in the domain M provided that $(R/2 - z + z_0) \gg 1/g\sqrt{\xi}$. Here σ is an arbitrary phase which cannot be gauged away. Inside the wall the second quark flavor tends to zero exponentially too.

It is not difficult to check that the main contribution to the wall tension comes from the middle layer while the edge domains produce contributions of the order of $\xi^{3/2}$ which makes them negligibly small.

Now let us comment on the phase factor in (8.2.9). Its origin is as follows [142]. The bulk theory at $\Delta m \neq 0$ has the U(1)×U(1) flavor symmetry corresponding to two independent rotations of two quark flavors. In both vacua only one quark develops a VEV. Therefore, in both vacua only one of these two U(1)'s is broken. The corresponding phase is eaten by the Higgs mechanism. However, on the wall both quarks have nonvanishing values, breaking both U(1) groups. Only one of the corresponding two phases is eaten by the Higgs mechanism. The other one becomes a Goldstone mode living on the wall.

Thus, we have two collective coordinates characterizing our wall solution, the position of the center z_0 and the phase σ. In the effective low-energy theory on the wall they become scalar fields of the world volume (2+1)-dimensional theory, $z_0(t, x, y)$ and $\sigma(t, x, y)$, respectively. The target space of the second field is S_1.

This wall is a 1/2 BPS solution of the Bogomol'nyi equations. In other words, four out of eight supersymmetry generators of the $\mathcal{N} = 2$ bulk theory are broken. As was shown in [142], the four supercharges selected by the conditions

$$\bar{\varepsilon}_{\dot{2}}^2 = -i\varepsilon^{21}, \quad \bar{\varepsilon}_{\dot{2}}^1 = -i\varepsilon^{22},$$
$$\bar{\varepsilon}_{\dot{1}}^1 = i\varepsilon^{12}, \quad \bar{\varepsilon}_{\dot{1}}^2 = i\varepsilon^{11}, \qquad (8.2.10)$$

act trivially on the wall solution. They become the four supersymmetries acting in the (2+1)-dimensional effective world volume theory on the wall. Here $\varepsilon^{\alpha f}$ and $\bar{\varepsilon}_{\dot{\alpha}}^f$ are eight supertransformation parameters.

8.3 Effective field theory on the wall

In this section we will review the (2+1)-dimensional effective low-energy theory of the moduli on the wall [142]. To this end we will make the wall collective coordinates z_0 and σ (together with their fermionic superpartners) slowly varying fields depending on x_n ($n = 0, 1, 2$), For simplicity let us consider the bosonic fields $z_0(x_n)$ and $\sigma(x_n)$; the residual supersymmetry will allow us to readily reconstruct the fermion part of the effective action.

Because $z_0(x_n)$ and $\sigma(x_n)$ correspond to zero modes of the wall, they have no potential terms in the world sheet theory. Therefore, in fact our task is to derive their kinetic terms, much in the same way as it was done for strings, see Section 4.4. For $z_0(x_n)$ this procedure is very simple. Substituting the wall solution (8.2.5), (8.2.8), and (8.2.9) in the action (8.1.1) and taking into account the x_n dependence of this modulus we immediately get

$$\frac{T_w}{2} \int d^3x \, (\partial_n z_0)^2. \qquad (8.3.1)$$

As far as the kinetic term for $\sigma(x_n)$ is concerned more effort is needed. We start from Eqs. (8.2.8) and (8.2.9) for the quark fields. Then we will have to modify our *ansatz* introducing nonvanishing components of the gauge field,

$$A_n = \chi(z)\,\partial_n\sigma(x_n). \tag{8.3.2}$$

These components of the gauge field are needed to make the world volume action well-defined. They are introduced in order to cancel the x dependence of the quark fields far away from the wall (in the quark vacua at $z \to \infty$) emerging through the x dependence of $\sigma(x_n)$, see Eq. (8.2.9).

Thus, we introduce a new profile function $\chi(z)$. It has no role in the construction of the static wall solution *per se*. It is unavoidable, however, in constructing the kinetic part of the world sheet theory of the moduli. This new profile function is described by its own action, which will be subject to minimization procedure. This is quite similar to derivation of the world sheet effective theory for non-Abelian strings, see Section 4.4.

The gauge potential in Eq. (8.3.2) is pure gauge far away from the wall and is not pure gauge inside the wall. It does lead to a nonvanishing field strength.

To ensure proper vacua at $z \to \pm\infty$ we impose the following boundary conditions on the function $\chi(z)$

$$\chi(z) \to 0, \quad z \to -\infty,$$
$$\chi(z) \to -2, \quad z \to +\infty. \tag{8.3.3}$$

Remember, the electric charge of the quark fields is $\pm 1/2$.

Next, substituting Eqs. (8.2.8), (8.2.9) and (8.3.2) in the action (8.1.1) we arrive at

$$S_{2+1}^{\sigma} = \left[\int d^3x \, \frac{1}{2}(\partial_n\sigma)^2 \right]$$
$$\times \int dz \left\{ \frac{1}{g^2}(\partial_z\chi)^2 + \chi^2|q^1|^2 + (2+\chi)^2|q^2|^2 \right\}. \tag{8.3.4}$$

The expression in the second line must be considered as an "action" for the χ profile function.

Our next task is to explicitly find the function χ. To this end we have to minimize (8.3.4) with respect to χ. This gives the following equation:

$$-\partial_z^2\chi + g^2\chi|q^1|^2 + g^2(2+\chi)|q^2|^2 = 0. \tag{8.3.5}$$

The equation for χ is of the second order. This is because the domain wall is no longer BPS state once we switch on the dependence of the moduli on the "longitudinal" variables x_n.

To the leading order in $g\sqrt{\xi}/\Delta m$ the solution of Eq. (8.3.5) can be obtained in the same manner as it was done previously for other profile functions. Let us first discuss what happens outside the inner part of the wall. Say, at $z - z_0 \gg R/2$ the profile $|q^1|$ vanishes while $|q^2|$ is exponentially close to $\sqrt{\xi}$ and, hence,

$$\chi \rightarrow -2 + \text{const } e^{-m_\gamma (z-z_0)}. \tag{8.3.6}$$

At $z_0 - z \gg R/2$ the profile function χ falls off exponentially to zero. Thus, outside the inner part of the wall, at $|z - z_0| \gg R/2$, the function χ approaches its boundary values with the exponential rate of approach.

Of most interest, however, is the inside part, the middle domain M (see Fig. 8.1). Here both quark profile functions vanish, and Eq. (8.3.5) degenerates into $\partial_z^2 \chi = 0$. As a result, the solution takes the form

$$\chi = -1 - 2\frac{z - z_0}{R}. \tag{8.3.7}$$

In the narrow edge domains $E_{1,2}$ the exact χ profile smoothly interpolates between the boundary values, see Eq. (8.3.6), and the linear behavior (8.3.7) inside the wall. These edge domains give small corrections to the leading term in the action.

Substituting the solution (8.3.7) in the χ action, the second line in Eq. (8.3.4), we finally arrive at

$$S_{2+1}^\sigma = \frac{\xi}{\Delta m} \int d^3x \, \frac{1}{2} (\partial_n \sigma)^2. \tag{8.3.8}$$

As well-known [227], the compact scalar field $\sigma(t, x, y)$ can be reinterpreted to be dual to the (2+1)-dimensional Abelian gauge field living on the wall. The emergence of the gauge field on the wall is easy to understand. The quark fields almost vanish inside the wall. Therefore the U(1) gauge group is restored inside the wall while it is Higgsed in the bulk. The dual U(1) is in the confinement regime in the bulk. Hence, the dual U(1) gauge field is localized on the wall, in full accordance with the general argument of Ref. [11]. The compact scalar field $\sigma(x_n)$ living on the wall is a manifestation of this magnetic localization.

The action in Eq. (8.3.8) implies that the coupling constant of our effective U(1) theory on the wall is given by

$$e^2 = 4\pi^2 \frac{\xi}{\Delta m}. \tag{8.3.9}$$

In particular, the definition of the (2+1)-dimensional gauge field takes the form

$$F_{nm}^{(2+1)} = \frac{e^2}{2\pi} \varepsilon_{nmk} \partial^k \sigma. \tag{8.3.10}$$

This finally leads us to the following effective low-energy theory of the moduli fields on the wall:

$$S_{2+1} = \int d^3x \left\{ \frac{T_w}{2} (\partial_n z_0)^2 + \frac{1}{4 e^2} (F_{nm}^{(2+1)})^2 + \text{fermion terms} \right\}. \quad (8.3.11)$$

The fermion content of the world volume theory is given by two three-dimensional Majorana spinors, as is required by $\mathcal{N} = 2$ in three dimensions (four supercharges, see (8.2.10)). The full world volume theory is a U(1) gauge theory in (2+1) dimensions, with four supercharges. The Lagrangian and the corresponding superalgebra can be obtained by reducing four-dimensional $\mathcal{N} = 1$ SQED (with no matter) to three dimensions.

The field z_0 in (8.3.11) is the $\mathcal{N} = 2$ superpartner of the gauge field A_n. To make it more transparent we make a rescaling, introducing a new field

$$a_{2+1} = 2\pi \xi \, z_0. \quad (8.3.12)$$

In terms of a_{2+1} the action (8.3.11) takes the form

$$S_{2+1} = \int d^3x \left\{ \frac{1}{2e^2} (\partial_n a_{2+1})^2 + \frac{1}{4e^2} (F_{mn}^{(2+1)})^2 + \text{fermions} \right\}. \quad (8.3.13)$$

The gauge coupling constant e^2 has dimension of mass in three dimensions. A characteristic scale of massive excitations on the world volume theory is of the order of the inverse thickness of the wall $1/R$, see (8.2.7). Thus, the dimensionless parameter that characterizes the coupling strength in the world volume theory is $e^2 R$,

$$e^2 R = \frac{16\pi^2}{g^2}. \quad (8.3.14)$$

This can be interpreted as a feature of the bulk–wall duality: the weak coupling regime in the bulk theory corresponds to strong coupling on the wall and *vice versa* [142, 228]. Of course, finding explicit domain wall solutions and deriving the effective theory on the wall assumes weak coupling in the bulk, $g^2 \ll 1$. In this limit the world volume theory is in the strong coupling regime and is not very useful.

The fact that each domain wall has two bosonic collective coordinates – its center and the phase – in the sigma model limit was noted in [214, 221].

To summarize, we showed that the world volume theory on the domain wall is the U(1) gauge theory (8.3.13) with extended supersymmetry, $\mathcal{N} = 2$. Thus, the domain wall in the theory (8.1.1) presents an example of a field-theoretic D brane: it localizes a gauge field on its world volume. In string theory gauge fields are localized on D branes because fundamental open strings can end on D branes. It turns out that this is also true for field-theoretic "*D* branes." In fact, various junctions of field-theoretic strings (flux tubes) with domain walls were found explicitly [215, 142, 37]. We will review 1/4-BPS junctions in Chapter 9. Meanwhile, in Section 8.4 we will consider non-Abelian generalizations of the localization effect for the gauge fields.

8.4 Domain walls in the U(N) gauge theories

In this section we will review the domain walls in $\mathcal{N} = 2$ SQCD (see Eq. (4.1.7)) with the U(N) gauge group. We assume that the number of the quark flavors in this theory $N_f > N$, so the theory has many vacua of the type (4.1.11), (4.1.14) depending on which N quarks out of N_f develop VEVs. We can denote different vacua as $(A_1, A_2, \ldots, A_N)$ specifying which quark flavors develop VEVs. Mostly, we will consider a general case assuming all quark masses to be different.

Let us arrange the quark masses as follows:

$$m_1 > m_2 > \cdots > m_{N_f}. \tag{8.4.1}$$

In this case the theory (4.1.7) has

$$\frac{N_f!}{N!(N_f - N)!} \tag{8.4.2}$$

isolated vacua.

Domain walls interpolating between these vacua were classified in [218]. Below we will briefly review this classification.

The Bogomol'nyi representation of the action (4.1.7) leads to the first-order equations for the wall configurations [229], see also [37],

$$\partial_z \varphi^A = -\frac{1}{\sqrt{2}} \left(a_a \tau^a + a + \sqrt{2} m_A \right) \varphi^A,$$

$$\partial_z a^a = -\frac{g_2^2}{2\sqrt{2}} \left(\bar{\varphi}_A \tau^a \varphi^A \right),$$

$$\partial_z a = -\frac{g_1^2}{2\sqrt{2}} \left(|\varphi^A|^2 - 2\xi \right), \tag{8.4.3}$$

where we used the *ansatz* (4.2.1) and introduced a single quark field φ^{kA} instead of two fields q^{kA} and $\tilde{q}_{Ak}$. These walls are 1/2 BPS saturated. The wall tensions are given by the surface term

$$T_w = \sqrt{2}\xi \int dz \, \partial_z a. \tag{8.4.4}$$

They can be written as [218]

$$T_w = \xi \, \vec{g} \, \vec{m},$$

(8.4.5)

where we use Eq. (4.1.11) and define $\vec{m} = (m_1, \ldots, m_{N_f})$, while

$$\vec{g} = \sum_{i=1}^{N_f-1} k_i \vec{\alpha}_i \, .$$

(8.4.6)

Here k_i are integers while α_i are simple roots of the U(N_f) algebra,[1]

$$\vec{\alpha}_1 = (1, -1, 0, \ldots, 0),$$
$$\vec{\alpha}_2 = (0, 1, -1, \ldots, 0),$$
$$\cdots ,$$
$$\vec{\alpha}_{N_f-1} = (0, \ldots, 0, 1, -1).$$

(8.4.7)

Elementary walls arise if one of the k_i's reduces to unity while all other integers in the set vanish. The tensions of the elementary walls are

$$T_w^i = \xi \, (m_i - m_{i+1}).$$

(8.4.8)

The ith elementary wall interpolates between the vacua $(\ldots, i, \ldots)$ and $(\ldots, i+1, \ldots)$. All other walls can be considered as composite states of elementary walls.

As an example let us consider the theory (4.1.7) with the gauge group U(2) and $N_f = 4$. Explicit solutions for the elementary walls in the limit

$$(m_i - m_{i+1}) \gg g\sqrt{\xi}$$

(8.4.9)

were obtained in [37]. They have the same three-layer structure as in the Abelian case, see Section 8.2. Say, the elementary wall interpolating between the vacua $(1, 2)$ and $(1, 3)$ has the following structure. At the left edge the quark φ^2 varies from its VEV $\sqrt{\xi}$ to zero exponentially, while at the right edge the quark φ^3 evolves from zero to its VEV $\sqrt{\xi}$. In the broad middle domain the fields a and a^3 linearly interpolate between their VEVs in two vacua. A novel feature of the domain wall solution as compared to the Abelian case (see Section 8.2) is that the quark field φ^1 does not vanish both outside and inside the wall.

---

[1] Each $\vec{\alpha}$ in Eq. (8.4.7) is an N_f-component vector, rather than $(N_f - 1)$-component vector of SU(N_f). The Cartan generators H_i ($i = 1, 2, \ldots, N_f$) are $N_f \times N_f$ diagonal matrices, $(H_i)_{kl} = \delta_{ki}\delta_{li}$, while the relevant non-Cartan generators $E_{\vec{\alpha}_i}$ are defined as $(E_{\vec{\alpha}_i})_{i,i+1} = 1$, with all other entries vanishing.

The solution for the elementary wall has two real moduli much in the same way as in the Abelian case: the wall center z_0 and a compact phase. The phase can be rewritten as a U(1) gauge field. Therefore, the effective theory on the elementary wall is of the type (8.3.13), as in the Abelian case. The physical reason behind the localization of the U(1) gauge field on the wall world volume is easy to understand. Since the quark φ^1 does not vanish inside the wall only an appropriately chosen U(1) field, namely $(A_\mu - A_\mu^3)$, which does not interact with this quark field can propagate freely inside the wall.

In the case of generic quark masses the effective world volume theory for composite domain walls contains U(1) gauge fields associated with each elementary wall. However, the metric on the moduli space can be more complicated. For example the metric for the $\vec{\alpha}_1 + \vec{\alpha}_2$ composite wall was shown [221, 230] to have a cigar-like geometry.

We conclude this section noting that the case of the degenerate quark masses was considered in [37, 219]. In particular, in [37] the $N = 2$ case was studied and it was argued that the composite wall made of two elementary walls localizes a non-Abelian U(2) gauge field. In [219] non-localized zero modes which were called "non-Abelian clouds" were found on the composite wall.

9

Wall-string junctions

In Chapter 8 we reviewed the construction of D-brane prototypes in field theory. In string theory D branes are extended objects on which fundamental strings can end. To make contact with this string/brane picture one may address a question whether or not solitonic strings can end on the domain wall which localizes gauge fields. The answer to this question is yes. Moreover, the string endpoint plays a role of a charge with respect to the gauge field localized on the wall surface. This issue was studied in [215] in the sigma-model setup and in [217] for gauge theories at strong coupling. A solution for a 1/4-BPS wall-string junction in the $\mathcal{N} = 2$ supersymmetric U(1) gauge theory at weak coupling was found in [142], while [37] deals with its non-Abelian generalization. Further studies of the wall-string junctions were carried out in [172] where all 1/4-BPS solutions to Eqs. (4.5.5) were obtained, and in [218, 231] where the energy associated with the wall-string junction (boojum) was calculated, and in [232, 228] where a quantum version of the effective theory on the domain wall world volume which takes into account charged matter (strings of the bulk theory) was derived. Below we will review the wall-string junction solutions and then briefly discuss how the presence of strings in the bulk modifies the effective theory on the wall.

9.1 Strings ending on the wall

To begin with, let us review the solution for the simplest 1/4-BPS wall-string junction in $\mathcal{N} = 2$ SQED obtained in [142]. As was discussed in Chapter 8, in both vacua of the theory the gauge field is Higgsed while it can spread freely inside the wall. This is the physical reason why the ANO string carrying a magnetic flux can end on the wall [11, 213].

Assume that at large distances from the string endpoint which lies at $r = 0$, $z = 0$ the wall is almost parallel to the (x_1, x_2) plane while the string is stretched along the z axis. As usual, we look for a static solution assuming that all relevant

209

fields can depend only on x_n, $(n = 1, 2, 3)$. The Abelian version of the first-order
equations (4.5.5) for various 1/4-BPS junctions in the theory (8.1.1) is [142]

$$F_1^* - iF_2^* - \sqrt{2}(\partial_1 - i\partial_2)a = 0,$$

$$F_3^* - \frac{g^2}{2}(|q^A|^2 - \xi) - \sqrt{2}\,\partial_3 a = 0,$$

$$\nabla_3 q^A = -\frac{1}{\sqrt{2}}q^A(a + \sqrt{2}m_A),$$

$$(\nabla_1 - i\nabla_2)q^A = 0. \tag{9.1.1}$$

These equations generalize the first-order equations for the wall (8.2.2) and the
Abelian ANO string.

Needless to say, the solution of the first-order equations (9.1.1) for a string ending
on the wall can be found only numerically especially near the endpoint of the string
where both the string and the wall profiles are heavily deformed. However, far away
from the string endpoint, deformations are weak and we can find the asymptotic
behavior analytically.

Let the string be on the $z > 0$ side of the wall, where the vacuum is given by
Eq. (8.1.5). First note that in the region $z \to \infty$ far away from the string endpoint
at $z \sim 0$ the solution to (9.1.1) is given by an almost unperturbed ANO string.
Now consider the domain $r \to \infty$ at small z. In this domain the solution to (9.1.1)
is given by a perturbation of the wall solution. Let us use the *ansatz* in which the
solutions for the fields a and q^A are given by the same equations (8.2.5), (8.2.8)
and (8.2.9) in which the size of the wall is still given by (8.2.7), and the only
modification is that the position of the wall z_0 and the phase σ now become slowly
varying functions of r and α, the polar coordinates on the (x_1, x_2) plane. It is quite
obvious that z_0 will depend only on r, as schematically depicted in Fig. 9.1.

Substituting this *ansatz* into the first-order equations (9.1.1) one arrives at the
equations which determine the adiabatic dependence of the moduli z_0 and σ on
r and α [142],

$$\partial_r z_0 = -\frac{1}{\Delta m r}, \tag{9.1.2}$$

$$\frac{\partial \sigma}{\partial \alpha} = 1, \quad \frac{\partial \sigma}{\partial r} = 0. \tag{9.1.3}$$

Needless to say our adiabatic approximation holds only provided the r derivative
is small, i.e. sufficiently far from the string, $\sqrt{\xi}r \gg 1$. The solution to Eq. (9.1.2)
is straightforward,

$$z_0 = -\frac{1}{\Delta m} \ln r + \text{const.} \tag{9.1.4}$$

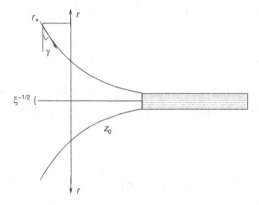

Figure 9.1. Bending of the wall due to the string-wall junction. The flux tube extends to the right infinity. The wall profile is logarithmic at transverse distances
larger than $\xi^{-1/2}$ from the string axis. At smaller distances the adiabatic
approximation fails.

We see that the wall is logarithmically bent according to the Coulomb law in $2+1$
dimensions (see Fig. 9.1). This bending produces a balance of forces between the
string and the wall in the z direction so that the whole configuration is static. The
solution to Eq. (9.1.3) is

$$\sigma = \alpha . \tag{9.1.5}$$

This vortex solution is certainly expected and welcome. One can identify the compact scalar field σ with the electric field living on the domain wall world volume
via (8.3.10). Equation (9.1.5) implies

$$F_{0i}^{(2+1)} = \frac{e^2}{2\pi} \frac{x_i}{r^2} \tag{9.1.6}$$

for this electric field, where the $(2+1)$-dimensional coupling is given by (8.3.9).

This is the field of a point-like electric charge in $2+1$ dimensions placed at
$x_i = 0$. The interpretation of this result is that the string endpoint on the wall plays
a role of the electric charge in the dual U(1) theory on the wall. From the standpoint
of the bulk theory, when the string ends on the wall, the magnetic flux it brings
with it spreads out inside the wall in accordance with the Coulomb law in $(2+1)$
dimensions.

From the above discussion it is clear that in the world volume theory (8.3.13),
the fields (9.1.4) and (9.1.6) can be considered as produced by classical point-like
charges which interact in a standard way with the electromagnetic field A_n and the
scalar field a_{2+1},

$$S_{2+1} = \int d^3x \left\{ \frac{1}{2e^2}(\partial_n a_{2+1})^2 + \frac{1}{4e^2}(F_{mn}^{(2+1)})^2 + A_n j_n - a_{2+1} \rho \right\}, \tag{9.1.7}$$

where the classical electromagnetic current and the charge density of static charges are

$$j_n(x) = n_e\{\delta^{(3)}(x), 0, 0\}, \quad \rho(x) = n_s \delta^{(3)}(x). \tag{9.1.8}$$

Here n_e and n_s are electric and scalar charges associated with the string endpoint with respect to the electromagnetic field A_n and the scalar field a, respectively [228],

$$
\begin{aligned}
n_e &= +1, \quad \text{incoming flux,} \\
n_e &= -1, \quad \text{outgoing flux,}
\end{aligned}
\tag{9.1.9}
$$

while their scalar charges are

$$
\begin{aligned}
n_s &= +1, \quad \text{string from the right,} \\
n_s &= -1, \quad \text{string from the left.}
\end{aligned}
\tag{9.1.10}
$$

These rules are quite obvious from the perspective of the bulk theory. The anti-string carries the opposite flux to that of a string in (9.1.6) and the bending of the wall produced by the string coming from the left is opposite to the one in (9.1.4) associated with the string coming from the right.

9.2 Boojum energy

Let us now calculate the energy of the wall-string junction, the boojum. There are two distinct contributions to this energy [231]. The first contribution is due to the gauge field (9.1.6),

$$
\begin{aligned}
E^G_{(2+1)} &= \int \frac{1}{2e^2_{2+1}} (F_{0i})^2 \, 2\pi r \, dr \\
&= \frac{\pi\xi}{\Delta m} \int \frac{dr}{r} = \frac{\pi\xi}{\Delta m} \ln\left(g\sqrt{\xi}L\right).
\end{aligned}
\tag{9.2.1}
$$

The integral $\int dr/r$ is logarithmically divergent both in the ultraviolet and infrared. It is clear that the UV divergence is cut off at the transverse size of the string $\sim 1/g\sqrt{\xi}$ and presents no problem. However, the infrared divergence is much more serious. We introduced a large size L to regularize it in (9.2.1).

The second contribution, due to the z_0 field (9.1.4), is proportional to $\int dr/r$ too,

$$
\begin{aligned}
E^H_{(2+1)} &= \int \frac{T_w}{2} (\partial_r z_0)^2 \, 2\pi r \, dr \\
&= \frac{\pi\xi}{\Delta m} \ln\left(g\sqrt{\xi}L\right).
\end{aligned}
\tag{9.2.2}
$$

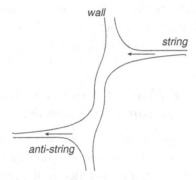

Figure 9.2. String and anti-string ending on the wall from different sides. Arrows denote the direction of the magnetic flux.

Both contributions are logarithmically divergent in the infrared. Their occurrence is an obvious feature of charged objects coupled to massless fields in $(2 + 1)$ dimensions due to the fact that the fields A_n and a_{2+1} do not die off at infinity, which means infinite energy.

The above two contributions are equal (with the logarithmic accuracy), even though their physical interpretation is different. The total energy of the string junction is

$$E^{G+H} = \frac{2\pi\xi}{\Delta m} \ln\left(g\sqrt{\xi}L\right). \tag{9.2.3}$$

We see that in our attempt to include strings as point-like charges in the world volume theory (9.1.7) we encounter problems already at the classical level. The energy of a single charge is IR divergent. It is clear that the infrared problems will become even more severe in quantum theory.

A way out was suggested in [231, 228]. For the infrared divergences to cancel we should consider strings and anti-strings with incoming and outgoing fluxes as well as strings coming from the right and from the left. Clearly, only configurations with vanishing total electric and scalar charges have finite energy (see (9.1.9) and (9.1.10)).

In fact, it was shown in [231] that the configuration depicted in Fig. 9.2 is a non-interacting 1/4-BPS configuration. All logarithmic contributions are canceled; the junction energy in this geometry is given by a finite negative contribution

$$E = -\frac{8\pi}{g^2}\Delta m, \tag{9.2.4}$$

which is called the boojum energy [218]. In fact this energy was first calculated in [172]. A procedure allowing one to separate this finite energy from logarithmic contributions described above (and make it well-defined) was discussed in [231].

9.3 Finite-size rigid strings stretched between the walls. Quantizing string endpoints

Now, after familiarizing ourselves with the junctions of the BPS walls with the semi-infinite strings, the boojums, we can ask whether or not the junction can acquire a dynamical role. Is there a formulation of the problem in which one can speak of a junction as of a particle sliding on the wall?

The string energy is its tension (4.2.12) times its length. If we have a single wall, all strings attached to it have half-infinite length; therefore, they are infinitely heavy. In the wall world volume theory (9.1.7) they may be seen as classical infinitely heavy point-like charges. The junctions are certainly non-dynamical objects in this case.

In order to be able to treat junctions as "particles" we need to make strings "light" and deprive them of their internal dynamics, i.e. switch off all string excitations. It turns out a domain in the parameter space is likely to exist where these goals can be achieved.

In this section we will review a quantum version of the world volume theory (9.1.7) with additional charged matter fields. The latter will represent the junctions on the wall world volume [232, 228] (of course, the junctions have strings of the bulk theory attached to them; these strings will be rigid).

Making string masses finite is a prerequisite. To this end one needs at least two domain walls at a finite distance from each other with strings stretched between them. This scenario was suggested in [232]. A quantum version of the wall world volume theory in which the strings were represented by a charged chiral matter superfield in $1 + 2$ dimensions was worked out. However, in the above scenario the strings were attached to each wall from one side. From the discussion in Section 9.2 it must be clear that this theory is not free from infrared problems. The masses of $(1 + 2)$-dimensional charged fields are infinite.

To avoid these infinities we need a configuration with strings coming both from the right and from the left sides of each wall. This configuration was suggested in [228], see Fig. 9.3.

Let us describe this set-up in more detail. First, we compactify the $x_3 = z$ direction in our bulk theory (8.1.1) on a circle of length L. Then we consider a pair "wall plus antiwall" oriented in the $\{x_1, x_2\}$ plane, separated by a distance l in the perpendicular direction, as shown in Fig. 9.3. The wall and antiwall experience attractive forces. Strictly speaking, this is not a BPS configuration –

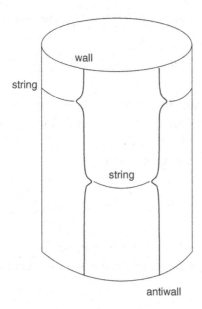

Figure 9.3. A wall and antiwall connected by strings on the cylinder. The circumference of the circle (the transverse slice of the cylinder) is L.

supersymmetry in the world volume theory is broken. However, the wall-antiwall interaction due to overlap of their profile functions is exponentially suppressed at large separations,

$$L \sim l \gg R, \qquad (9.3.1)$$

where R is the wall thickness (see Eq. (8.2.7)). In what follows we will neglect exponentially suppressed effects. If so, we neglect the effects which break super-symmetry in our $(2+1)$-dimensional world volume theory. Thus, it continues to have four conserved supercharges ($\mathcal{N} = 2$ supersymmetry in $(2+1)$ dimensions) as was the case for the isolated single wall.

Let us denote the wall position as z_1 while that of the antiwall as z_2. Then

$$l = z_2 - z_1.$$

On the wall world volume $z_{1,2}$ become scalar fields. The kinetic terms for these fields in the world volume theory are obvious (see (8.3.11)),

$$\frac{T_{\rm w}}{2}\left[(\partial_n z_1)^2 + (\partial_n z_2)^2\right] = \frac{1}{2e^2}\left[\left(\partial_n a^{(1)}_{2+1}\right)^2 + \left(\partial_n a^{(2)}_{2+1}\right)^2\right], \qquad (9.3.2)$$

where we use (8.3.12) to define the fields $a_{2+1}^{(1,2)}$. The sum of these fields,

$$a_+ \equiv \frac{1}{\sqrt{2}}\left(a_{2+1}^{(2)} + a_{2+1}^{(1)}\right),$$

with the corresponding superpartners, decouples from other fields forming a free field theory describing dynamics of the center-of-mass of our construction. This is a trivial part which will not concern us here.

An interesting part is associated with the field

$$a_- \equiv \frac{1}{\sqrt{2}}\left(a_{2+1}^{(2)} - a_{2+1}^{(1)}\right). \tag{9.3.3}$$

The factor $1/\sqrt{2}$ ensures that a_- has a canonically normalized kinetic term. By definition, it is related to the relative wall-antiwall separation, namely,

$$a_- = \frac{2\pi\xi}{\sqrt{2}}\, l\,. \tag{9.3.4}$$

Needless to say, a_- has all necessary $\mathcal{N} = 2$ superpartners. In the bosonic sector we introduce the gauge field

$$A_n^- \equiv \frac{1}{\sqrt{2}}\left(A_n^{(1)} - A_n^{(2)}\right), \tag{9.3.5}$$

with the canonically normalized kinetic term. The strings stretched between the wall and antiwall, on both sides, will be represented by two chiral superfields, S and $\tilde{S}$, respectively. We will denote the corresponding bosonic components by s and $\tilde{s}$.

In terms of these fields the quantum version of the theory (9.1.7) is completely determined by the charge assignments (9.1.9) and (9.1.10) and $\mathcal{N} = 2$ supersymmetry. The charged matter fields have the opposite electric charges and distinct mass terms, see below. A mass term for one of them is introduced by virtue of a "real mass," a three-dimensional generalization [96] of the twisted mass in two dimensions [32]. It is necessary due to the fact that there are two inter-wall distances, l and $L - l$. The real mass breaks parity. The bosonic part of the action has the form

$$S_{\text{bos}} = \int d^3x \left\{ \frac{1}{4e^2} F_{mn}^- F_{mn}^- + \frac{1}{2e^2}\left(\partial_n a_-\right)^2 + |\mathcal{D}_n s|^2 + |\tilde{\mathcal{D}}_n \tilde{s}|^2 \right.$$
$$\left. + 2a_-^2 \bar{\tilde{s}}\, s + 2(m - a_-)^2 \bar{\tilde{s}}\, \tilde{s} + e^2\left(|s|^2 - |\tilde{s}|^2\right)^2 \right\}. \tag{9.3.6}$$

According to our discussion in Section 9.1, the fields s and $\tilde{s}$ have charges $+1$ and -1 with respect to the gauge fields $A_n^{(1)}$ and $A_n^{(2)}$, respectively. Hence,

$$
\begin{aligned}
\mathcal{D}_n &= \partial_n - i\left(A_n^{(1)} - A_n^{(2)}\right) = \partial_n - i\sqrt{2}A_n^-, \\
\tilde{\mathcal{D}}_n &= \partial_n + i\left(A_n^{(1)} - A_n^{(2)}\right) = \partial_n + i\sqrt{2}A_n^-.
\end{aligned}
\tag{9.3.7}
$$

The electric charges of boojums with respect to the field A_n^- are $\pm\sqrt{2}$. The last term in (9.3.6) is the D term dictated by supersymmetry.

So far, m is a free parameter whose relation to L will be determined shortly. Moreover, $F_{mn}^- = \partial_m A_n^- - \partial_n A_m^-$. The theory (9.3.6) with the pair of chiral multiplets S and $\tilde{S}$ is free of IR divergences and global Z_2 anomalies [96, 98] (see also Section 3.2.1). At the classical level it is clear from our discussion in Section 9.2. A version of the world volume theory (9.3.6) with a single supermultiplet S was considered in Ref. [232] but, as was mentioned, this version is not free of IR divergences.

It is in order to perform a crucial test of our theory (9.3.6) by calculating the masses of the charged matter multiplets S and $\tilde{S}$. From (9.3.6) we see that the mass of S is

$$
m_s = \sqrt{2}\,\langle a_-\rangle.
\tag{9.3.8}
$$

Substituting here the relation (9.3.4) we get

$$
m_s = 2\pi\xi\,l.
\tag{9.3.9}
$$

The mass of the charged matter field S reduces to the mass of the string of the bulk theory stretched between the wall and antiwall at separation l, see (4.2.12). A great success! Of course, this was expected. Note that this is a nontrivial check of consistency between the world volume theory and the bulk theory. Indeed, the charges of the strings' endpoints (9.1.9) and (9.1.10) are unambiguously fixed by the classical solution for the wall-string junction.

Now, imposing the relation between the free mass parameter m in (9.3.6) and the length of the compactified z-direction L in the form

$$
m = \frac{2\pi\xi}{\sqrt{2}}\,L
\tag{9.3.10}
$$

we get the mass of the chiral field $\tilde{S}$ to be

$$
m_{\tilde{s}} = 2\pi\xi\,(L - l).
\tag{9.3.11}
$$

Figure 9.4. Mass scales of the bulk and world volume theories.

The mass of the string $\tilde{S}$ connecting the wall with the antiwall from the other side of the cylinder is the string tension times $(L - l)$, in full accordance with our expectations, see Fig. 9.3.

The theory (9.3.6) can be considered as an effective low-energy $(2 + 1)$-dimensional description of the wall-antiwall system dual to the $(3+1)$-dimensional bulk theory (8.1.1) under the choice of parameters specified below (Fig. 9.4). Most importantly, we use the quasiclassical approximation in our bulk theory (8.1.1) to find the solution for the string-wall junction [142] and derive the wall-antiwall world volume effective theory (9.3.6). This assumes weak coupling in the bulk, $g^2 \ll 1$. According to the duality relation (8.3.14) this implies strong coupling in the world volume theory.

In order to be able to work with the world volume theory we want to continue the theory (9.3.6) to the weak coupling regime,

$$ e^2 \ll \frac{1}{R}, \tag{9.3.12} $$

which means strong coupling in the bulk theory, $g^2 \gg 1$. The general idea is that at $g^2 \ll 1$ we can use the bulk theory (8.1.1) to describe our wall-antiwall system while at $g^2 \gg 1$ we better use the world volume theory (9.3.6). In [228] this set-up was termed bulk–brane duality. In spirit – albeit not in detail – it is similar to the AdS/CFT correspondence.

In order for the theory (9.3.6) to give a correct low-energy description of the wall-antiwall system the masses of strings (including boojums) in this theory should be much less than the masses of both the wall and string excitations. These masses are of order of $1/R$ and $m_{KK} = k/l \sim k/(L - l)$, respectively, where k is an integer. The high mass gap for the string excitations make strings rigid.

These constraints were studied in [228]. It was found that for the constraints to be satisfied different scales of the theory must have a hierarchy shown in Fig. 9.4.

The scales Δm, $\sqrt{\xi}$ and $e_{2+1}^2 \sim \xi/\Delta m$ are determined by the string and wall tensions in our bulk theory, see (4.2.12) and (8.2.4). In particular, the $(2 + 1)$-dimensional coupling e^2 is determined by the ratio of the wall tension to the square of the string tension, as follows from Eqs. (8.3.11) and (8.3.12). Since the strings and walls in the bulk theory are BPS-saturated, they receive no quantum corrections.

Equations (4.2.12) and (8.2.4) can be continued to the strong coupling regime in the bulk theory. Therefore, we can always take such values of the parameters Δm and $\sqrt{\xi}$ that the conditions

$$e^2 \ll \sqrt{\xi} \ll \Delta m \qquad (9.3.13)$$

are satisfied.

To actually prove duality between the bulk theory (8.1.1) and the world volume theory (9.3.6) we only need to prove the condition

$$\sqrt{\xi} \ll \frac{1}{R}, \qquad (9.3.14)$$

which ensures that strings are lighter than the wall excitations. This will give us the hierarchy of the mass scales shown in Fig. 9.4. With the given values of the parameters Δm and $\sqrt{\xi}$ we have another free parameter of the bulk theory to ensure (9.3.14), namely, the coupling constant g^2. However, the scale $1/R$ (the mass scale of various massive excitations of the wall) is not protected by supersymmetry and we cannot prove that the regime (9.3.14) can be reached at strong coupling in the bulk theory. Thus, the above bulk–brane duality is in fact a conjecture essentially equivalent to the statement that the regime (9.3.14) is attainable under a certain choice of parameters. Note, that if the condition (9.3.14) is not met, the wall excitations become lighter than the strings under consideration, and the theory (9.3.6) does not correctly describe low-energy physics of the theory on the walls.

9.4 Quantum boojums. Physics of the world volume theory

What is a boojum loop?

Let us integrate out the string multiplets S and $\tilde{S}$ and study the effective theory for the U(1) gauge supermultiplet at scales below m_s. As long as the string fields enter the action quadratically (if we do not resolve the algebraic equations for the auxiliary fields) the one-loop approximation is exact.

Integration over the charged matter fields in (9.3.6) leads to generation of the Chern–Simons term [94, 95, 96] with the coefficient proportional to

$$\frac{1}{4\pi}[\text{sign}(a) + \text{sign}(m - a)]\varepsilon_{nmk} A_n^- \partial_m A_k^-. \tag{9.4.1}$$

Another effect related to the one in (9.4.1) by supersymmetry is generation of a nonvanishing D-term,

$$\frac{D}{2\pi}[|m - a_-| - |a_-|] = \frac{D}{2\pi}(m - 2a_-), \tag{9.4.2}$$

where D is the D-component of the gauge supermultiplet. As a result we get from (9.3.6) the following low-energy effective action for the gauge multiplet:

$$S_{2+1} = \int d^3x \left\{ \frac{1}{2e^2(a_-)}(\partial_n a_-)^2 + \frac{1}{4e^2(a_-)}(F_{mn}^-)^2 \right.$$
$$\left. + \frac{1}{2\pi}\varepsilon_{nmk} A_n^- \partial_m A_k^- + \frac{e^2(a_-)}{8\pi^2}(2a_- - m)^2 \right\}, \tag{9.4.3}$$

where we also take into account a finite renormalization of the bare coupling constant e^2 [233, 234, 98],

$$\frac{1}{e^2(a_-)} = \frac{1}{e^2} + \frac{1}{8\pi |a_-|} + \frac{1}{8\pi |m - a_-|}. \tag{9.4.4}$$

This is a small effect since $1/e^2$ is the largest parameter (see Fig. 9.4). Note that in Eq. (9.4.3) the coefficient in front of the Chern–Simons term is an integer number (in the units of $1/(2\pi)$), as required by gauge invariance.

The most dramatic effect in (9.4.3) is the generation of a potential for the field a_-. Remember a_- is proportional to the separation l between the walls. The vacuum of (9.4.3) is located at

$$\langle a_- \rangle = \frac{m}{2}, \quad l = \frac{L}{2}. \tag{9.4.5}$$

There are two extra solutions at $a_- = 0$ and $a_- = m$, but they lie outside the limits of applicability of our approach.

We see that the wall and antiwall are pulled apart; they want to be located at the opposite sides of the cylinder. Moreover, the potential is quadratically rising with the deviation from the equilibrium point (9.4.5). As was mentioned in the beginning

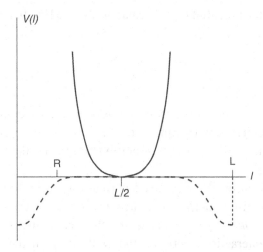

Figure 9.5. Classical and quantum wall-antiwall interaction potential. The dashed line depicts the classical exponentially small potential while the solid line the quantum potential presented in Eq. (9.4.3).

of this section, the wall and antiwall interact with exponentially small potential due to the overlap of their profiles. However, these interactions are negligibly small at $l \gg R$ as compared to the interaction in Eq. (9.4.3). The interaction potential in (9.4.3) arises due to virtual pairs of strings ("boojum loops") and it pulls the walls apart.

Clearly, our description of strings in the bulk theory was purely classical and we were unable to see this quantum effect. The classical and quantum interaction potential of the wall-antiwall system is schematically shown in Fig. 9.5. The quantum potential induced by virtual string loop is much larger than the classical exponentially small $W\bar{W}$ attraction at separations $l \sim L/2$. The quantum effect stabilizes the classically unstable $W\bar{W}$ system at the equilibrium position (9.4.5).

Note, that if the wall-antiwall interactions were mediated by particles they would have exponential fall-off at large separations l (there are no massless particles in the bulk). Quadratically rising potential would never be generated. In our case the interactions are due to virtual pairs of extended objects – strings. Strings are produced as rigid objects stretched between walls. The string excitations are not taken into account as they are too heavy. The fact that the strings come out in our treatment as rigid objects rather than local particle-like states propagating between the walls is of a paramount importance. This is the reason why the wall-antiwall potential does not fall off at large separations. Note that a similar effect, power-law interactions between the domain walls in $\mathcal{N} = 1$ SQCD, was obtained via a two-loop calculation in the effective world volume theory [235].

The presence of the potential for the scalar field a_- in Eq. (9.4.3) makes this field massive, with mass

$$m_a = \frac{e^2}{\pi}. \tag{9.4.6}$$

By supersymmetry, the photon is no longer massless too, it acquires the same mass. This is associated with the Chern–Simons term in (9.4.3). As it is clear from Fig. 9.4, $m_a \ll m_s$. This shows that integrating out massive string fields in (9.3.6) to get (9.4.3) makes sense.

Another effect seen in (9.4.3) is the renormalization of the coupling constant which results in a non-flat metric. Of course, this effect is very small in our range of parameters since $m_s \gg e^2$. Still we see that the virtual string pairs induce additional power interactions between the walls through the nontrivial metric in (9.4.3).

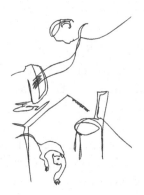

10

Conclusions

This concludes our travel diary in the land of supersymmetric solitons in gauge theories. It is time to summarize the lessons.

Advances in supersymmetric solitons, especially in non-Abelian gauge theories, that have taken place since 1996, are impressive. In the bulk of this book we thoroughly discussed many aspects of the subject at a technical level. Important and relevant technical details presented above should not overshadow the big picture, which has been in the making since 1973. Sometimes people tend to forget about this big picture which is understandable: its development is painfully slow and notoriously difficult.

Let us ask ourselves: what is the most remarkable feature of quantum chromodynamics and QCD-like theories? The fact that at the Lagrangian level one deals with quarks and gluons while experimentalists detect pions, protons, glueballs and other color singlet states – never quarks and gluons – is the single most salient feature of non-Abelian gauge theories at strong coupling. Color confinement makes colored degrees of freedom inseparable. In a bid to understand this phenomenon Nambu, 't Hooft and Mandelstam suggested in the mid 1970s (independently but practically simultaneously) a "non-Abelian dual Meissner effect." At that time their suggestion was more of a dream than a physical scenario. According to their vision, "non-Abelian monopoles" condense in the vacuum resulting in formation of "non-Abelian chromoelectric flux tubes" between color charges, e.g. between a probe heavy quark and antiquark. Attempts to separate these probe quarks would lead to stretching of the flux tubes, so that the energy of the system grows linearly with separation. That's how linear confinement was visualized. However, at that time the notions of non-Abelian flux tubes and non-Abelian monopoles (let alone condensed monopoles in non-Abelian gauge theories) were nonexistent. Nambu, 't Hooft and Mandelstam operated with nonexistent objects.

One may ask where these theorists got their inspiration from. There was one physical phenomenon known since long ago and well understood theoretically which yielded a rather analogous picture.

In 1933 Meissner discovered that magnetic fields could not penetrate inside superconducting media. The expulsion of the magnetic fields by superconductors goes under the name of the Meissner effect. Twenty years later Abrikosov posed the question: "What happens if one immerses a magnetic charge and an anticharge in type-II superconductors [which in fact he discovered]?" One can visualize a magnetic charge as an endpoint of a very long and very thin solenoid. Let us refer to the N endpoint of such a solenoid as a positive magnetic charge and the S endpoint as a negative magnetic charge.

In the empty space the magnetic field will spread in the bulk, while the energy of the magnetic charge-anticharge configuration will obey the Coulomb $1/r$ law. The force between them will die off as $1/r^2$.

What changes if the magnetic charges are placed inside a large type-II superconductor?

Inside the superconductor the Cooper pairs condense, all electric charges are screened, while the photon acquires a mass. According to modern terminology, the electromagnetic U(1) gauge symmetry is Higgsed. The magnetic field cannot be screened in this way; in fact, the magnetic flux is conserved. At the same time the superconducting medium does not tolerate the magnetic field.

The clash of contradictory requirements is solved through a compromise. A thin tube is formed between the magnetic charge and anticharge immersed in the superconducting medium. Inside this tube superconductivity is ruined – which allows the magnetic field to spread from the charge to the anticharge through this tube. The tube transverse size is proportional to the inverse photon mass while its tension is proportional to the Cooper pair condensate. These tubes go under the name of the Abrikosov vortices. In fact, for arbitrary magnetic fields he predicted lattices of such flux tubes. A dramatic (and, sometimes, tragic) history of this discovery is nicely described in Abrikosov's Nobel Lecture.

Returning to the magnetic charges immersed in the type-II superconductor under consideration, one can see that increasing the distance between these charges (as long as they are inside the superconductor) does not lead to their decoupling – the magnetic flux tubes become longer, leading to a linear growth of the energy of the system.

The Abrikosov vortex lattices were experimentally observed in the 1960s. This physical phenomenon inspired Nambu, 't Hooft and Mandelstam's ideas on non-Abelian confinement. Many people tried to quantify these ideas. The first breakthrough, instrumental in all current developments, came 20 years later, in the form of the Seiberg–Witten solution of $\mathcal{N} = 2$ super-Yang–Mills. This theory has eight supercharges which makes dynamics quite "rigid" and helps one to find the full analytic solution at low energies. The theory bears a resemblance to quantum chromodynamics, sharing common "family traits." By and large, one can characterize it as QCD's "second cousin."

An important feature which distinguishes it from QCD is the adjoint scalar field whose vacuum expectation value triggers the spontaneous breaking of the gauge symmetry $SU(2) \rightarrow U(1)$. The 't Hooft–Polyakov monopoles ensue. They are readily seen in the quasiclassical domain. Extended supersymmetry and holomorphy in certain parameters which is associated with it allows one to analytically continue in the domain where the monopoles become light – eventually massless – and then condense after a certain small deformation breaking $\mathcal{N} = 2$ down to $\mathcal{N} = 1$ is introduced. After that, at a much lower scale the (dual) $U(1)$ gauge symmetry breaks, so that the theory is fully Higgsed. Electric flux tubes are formed.

This was the first ever demonstration of the dual Meissner effect in non-Abelian theory, a celebrated analytic proof of linear confinement, which caused much excitement and euphoria in the community.

It took people three years to realize that the flux tubes in the Seiberg–Witten solution are not those we would like to have in QCD.[1] Hanany, Strassler and Zaffaroni, who analyzed the chromoelectric flux tubes in the Seiberg–Witten solution in 1997, showed that these flux tubes are essentially Abelian (of the Abrikosov–Nielsen–Olesen type) so that the hadrons they would create would not have much in common with those in QCD. The hadronic spectrum would be significantly richer. And, say, in the $SU(3)$ case, three flux tubes in the Seiberg–Witten solution would not annihilate into nothing, as they should in QCD ...

Ever since, searches for non-Abelian flux tubes and non-Abelian monopoles continued, with a decisive breakthrough in 2003. By that time the program of finding field-theory analogs of all basic constructions of string/D-brane theory was in full swing. BPS domain walls, analogs of D branes, had been identified in supersymmetric Yang–Mills theory. It had been demonstrated that such walls support gauge fields localized on them. BPS-saturated string-wall junctions had been constructed. And yet, non-Abelian flux tubes, the basic element of the non-Abelian Meissner effect, remained elusive.

They were first found in $U(2)$ super-Yang–Mills theories with extended supersymmetry, $\mathcal{N} = 2$, and two matter hypermultiplets. If one introduces a non-vanishing Fayet–Iliopoulos parameter ξ the theory develops isolated quark vacua, in which the gauge symmetry is fully Higgsed, and all elementary excitations are

---

[1] The Seiberg–Witten strings hopefully belong to the same universality class as the QCD strings, but this is impossible to prove with existing knowledge.

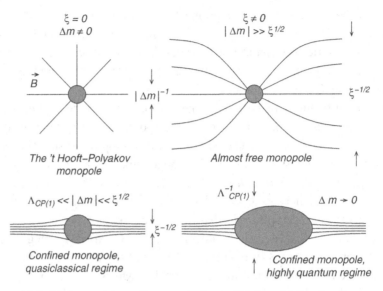

Figure 10.1. Various regimes for monopoles and strings in the simplest case of two flavors.

massive. In the general case, two matter mass terms allowed by $\mathcal{N} = 2$ are unequal, $m_1 \neq m_2$. There are free parameters whose interplay determines dynamics of the theory: the Fayet–Iliopoulos parameter ξ, the mass difference Δm and a dynamical scale parameter Λ, an analog of the QCD scale Λ_{QCD}. Extended supersymmetry guarantees that some crucial dependences are holomorphic, and there is no phase transition.

As various parameters vary, this theory evolves in a very graphic way, see Fig. 10.1 which is almost the same as Fig. 4.3 (the first stage of unconfined 't Hooft–Polyakov monopole is added in the left upper corner). At $\xi = 0$ but $\Delta m \neq 0$ (and $\Delta m \gg \Lambda$) it presents a very clear-cut example of a model with the standard 't Hooft–Polyakov monopole. The monopole is free to fly – the flux tubes are not yet formed.

Switching on $\xi \neq 0$ traps the magnetic fields inside the flux tubes, which are weak as long as $\xi \ll \Delta m$. The flux tubes change the shape of the monopole far away from its core, leaving the core essentially intact. Orientation of the chromomagnetic field inside the flux tube is essentially fixed. The flux tubes are Abelian.

With $|\Delta m|$ decreasing, fluctuations in the orientation of the chromomagnetic field inside the flux tubes grow. Simultaneously, the monopole seen as the string junction, loses resemblance with the 't Hooft–Polyakov monopole. It acquires a life of its own.

Finally, in the limit $\Delta m \to 0$ the transformation is complete. A global SU(2) symmetry restores in the bulk. Orientational moduli develop on the string world sheet making it non-Abelian. The junctions of degenerate strings present what remains of the monopoles in this highly quantum regime. It is remarkable that, despite the fact we are deep inside the highly quantum regime, holomorphy allows one to exactly calculate the mass of these monopoles.

What remains to be done? The most recent investigations zero in on $\mathcal{N} = 1$ theories, which are much closer relatives of QCD than $\mathcal{N} = 2$.

And then, $\mathcal{N} = 0$ theories – sister theories of QCD – loom large ...

Appendix A

Conventions and notation

The conventions we use in Parts I and II are slightly different. In Part I presenting mainly a conceptual introduction to the subject of supersymmetric solitons we choose the so-called Minkowski notation. Here our notation is very close (but not identical!) to that of Bagger and Wess [40]. The main distinction is the conventional choice of the metric tensor $g_{\mu\nu} = \text{diag}(+---)$ as opposed to the $\text{diag}(-+++)$ version of Bagger and Wess. Both the spinorial and vectorial indices will be denoted by Greek letters. To differentiate between them we will use the letters from the beginning of the alphabet for the spinorial indices, e.g. α, β and so on, reserving those from the end of the alphabet (e.g. μ, ν, *etc.*) for the vectorial indices.

Those readers who venture to delve into Part II will have to switch to the so-called Euclidean notation which is more convenient for technical studies. The distinctions between these two notations are summarized in Section A.7.

A.1 Two-dimensional gamma matrices

In two dimensions we choose the gamma matrices as follows

$$\gamma^0 = \gamma^t = \sigma_2, \quad \gamma^1 = \gamma^z = i\sigma_1, \quad \gamma^5 \equiv \gamma^0\gamma^1 = \sigma_3. \tag{A.1}$$

In three dimensions

$$\gamma^t = \sigma_2, \quad \gamma^x = -i\sigma_3, \quad \gamma^z = i\sigma_1. \tag{A.2}$$

A.2 Covariant derivatives

The covariant derivative in the Minkowski space is defined as

$$D_\mu = \partial_\mu - iA_\mu^a T^a. \tag{A.3}$$

Then for the spatial derivatives we have

$$D_1 = \frac{\partial}{\partial x} + i A_x^a T^a, \tag{A.4}$$

and similar expressions for $D_{2,3}$.

A.3 Superspace and superfields

The four-dimensional space x^μ can be promoted to superspace by adding four Grassmann coordinates θ_α and $\bar\theta_{\dot\alpha}$, $(\alpha, \dot\alpha = 1, 2)$. The coordinate transformations

$$\{x^\mu, \theta_\alpha, \bar\theta_{\dot\alpha}\}: \quad \delta\theta_\alpha = \varepsilon_\alpha, \quad \delta\bar\theta_{\dot\alpha} = \bar\varepsilon_{\dot\alpha}, \quad \delta x_{\alpha\dot\alpha} = -2i\,\theta_\alpha\bar\varepsilon_{\dot\alpha} - 2i\,\bar\theta_{\dot\alpha}\varepsilon_\alpha \tag{A.5}$$

add SUSY to the translational and Lorentz transformations.

Here the Lorentz vectorial indices are transformed into spinorial according to the standard rule

$$A_{\beta\dot\beta} = A_\mu(\sigma^\mu)_{\beta\dot\beta}, \quad A^\mu = \frac{1}{2}A_{\alpha\dot\beta}(\bar\sigma^\mu)^{\dot\beta\alpha}, \tag{A.6}$$

where

$$(\sigma^\mu)_{\alpha\dot\beta} = \{1, \vec\tau\}_{\alpha\dot\beta}, \quad (\bar\sigma^\mu)_{\dot\beta\alpha} = (\sigma^\mu)_{\alpha\dot\beta}. \tag{A.7}$$

We use the notation $\vec\tau$ for the Pauli matrices throughout the book. The lowering and raising of the indices is performed by virtue of the $\epsilon^{\alpha\beta}$ symbol ($\varepsilon^{\alpha\beta} = i(\tau_2)_{\alpha\beta}$). For instance,

$$(\bar\sigma^\mu)^{\dot\beta\alpha} = \epsilon^{\dot\beta\dot\rho}\,\epsilon^{\alpha\gamma}\,(\bar\sigma^\mu)_{\dot\rho\gamma} = \{1, -\vec\tau\}_{\dot\beta\alpha}. \tag{A.8}$$

Note that

$$\varepsilon^{12} = -\varepsilon_{12} = 1, \tag{A.9}$$

and the same for the dotted indices.

Two invariant subspaces $\{x_L^\mu, \theta_\alpha\}$ and $\{x_R^\mu, \bar\theta_{\dot\alpha}\}$ are spanned on 1/2 of the Grassmann coordinates,

$$\{x_L^\mu, \theta_\alpha\}: \quad \delta\theta_\alpha = \varepsilon_\alpha, \quad \delta(x_L)_{\alpha\dot\alpha} = -4i\,\theta_\alpha\bar\varepsilon_{\dot\alpha};$$

$$\{x_R^\mu, \bar\theta_{\dot\alpha}\}: \quad \delta\bar\theta_{\dot\alpha} = \bar\varepsilon_{\dot\alpha}, \quad \delta(x_R)_{\alpha\dot\alpha} = -4i\,\bar\theta_{\dot\alpha}\varepsilon_\alpha, \tag{A.10}$$

where

$$(x_{L,R})_{\alpha\dot\alpha} = x_{\alpha\dot\alpha} \mp 2i\,\theta_\alpha\bar\theta_{\dot\alpha}. \tag{A.11}$$

The minimal supermultiplet of fields includes one complex scalar field $\phi(x)$ (two bosonic states) and one complex Weyl spinor $\psi^\alpha(x)$, $\alpha = 1, 2$ (two fermionic states). Both fields are united in one *chiral superfield*,

$$\Phi(x_L, \theta) = \phi(x_L) + \sqrt{2}\theta^\alpha \psi_\alpha(x_L) + \theta^2 F(x_L), \tag{A.12}$$

where F is an auxiliary component. The field F appears in the Lagrangian without the kinetic term.

In the gauge theories one also uses a *vector superfield*,

$$\begin{aligned}
V(x, \theta, \bar\theta) = & C + i\theta\chi - i\bar\theta\bar\chi + \frac{i}{\sqrt{2}}\theta^2 M - \frac{i}{\sqrt{2}}\bar\theta^2 \bar M \\
& - 2\theta_\alpha\bar\theta_{\dot\alpha} v^{\dot\alpha\alpha} + \left\{ 2i\theta^2\bar\theta_{\dot\alpha} \left[\bar\lambda^{\dot\alpha} - \frac{i}{4}\partial^{\alpha\dot\alpha}\chi \right] + \text{H.c.} \right\} \\
& + \theta^2\bar\theta^2 \left[D - \frac{1}{4}\partial^2 C \right].
\end{aligned} \tag{A.13}$$

The superfield V is real, $V = V^\dagger$, implying that the bosonic fields C, D and $v^\mu = \sigma^\mu_{\alpha\dot\alpha} v^{\dot\alpha\alpha}/2$ are real. Other fields are complex, and the bar denotes the complex conjugation. The field strength superfield has the form

$$W_\alpha = i\left(\lambda_\alpha + i\theta_\alpha D - \theta^\beta F_{\alpha\beta} - i\theta^2 D_{\alpha\dot\alpha}\bar\lambda^{\dot\alpha} \right). \tag{A.14}$$

The gauge field strength tensor is denoted by $F^a_{\mu\nu}$. Sometimes we use the abbreviation F^2 for

$$F^2 \equiv F^a_{\mu\nu} F^{\mu\nu\,a}, \tag{A.15}$$

while

$$FF^* \equiv F^a_{\mu\nu} F^{*\mu\nu\,a} \equiv \frac{1}{2}\epsilon^{\mu\nu\rho\sigma} F^a_{\mu\nu} F^a_{\rho\sigma}. \tag{A.16}$$

The transformations (A.10) generate the SUSY transformations of the fields which can be written as

$$\delta V = i\left(Q\varepsilon + \bar Q\bar\varepsilon \right) V \tag{A.17}$$

where V is a generic superfield (which could be chiral as well). The differential operators Q and $\bar Q$ can be written as

$$Q_\alpha = -i\frac{\partial}{\partial\theta^\alpha} + \partial_{\alpha\dot\alpha}\bar\theta^{\dot\alpha}, \quad \bar Q_{\dot\alpha} = i\frac{\partial}{\partial\bar\theta^{\dot\alpha}} - \theta^\alpha\partial_{\alpha\dot\alpha}, \quad \{Q_\alpha, \bar Q_{\dot\alpha}\} = 2i\partial_{\alpha\dot\alpha}. \tag{A.18}$$

These differential operators give the explicit realization of the SUSY algebra,

$$\{Q_\alpha, \bar{Q}_{\dot{\alpha}}\} = 2P_{\alpha\dot{\alpha}}, \quad \{Q_\alpha, Q_\beta\} = 0, \quad \{\bar{Q}_{\dot{\alpha}}, \bar{Q}_{\dot{\beta}}\} = 0, \quad [Q_\alpha, P_{\beta\dot{\beta}}] = 0,$$
(A.19)

where Q_α and $\bar{Q}_{\dot{\alpha}}$ are the supercharges while $P_{\alpha\dot{\alpha}} = i\partial_{\alpha\dot{\alpha}}$ is the energy-momentum operator. The *superderivatives* are defined as the differential operators $\bar{D}_\alpha$, $D_{\dot{\alpha}}$ anticommuting with Q_α and $\bar{Q}_{\dot{\alpha}}$,

$$D_\alpha = \frac{\partial}{\partial\theta^\alpha} - i\partial_{\alpha\dot{\alpha}}\bar{\theta}^{\dot{\alpha}}, \quad \bar{D}_{\dot{\alpha}} = -\frac{\partial}{\partial\bar{\theta}^{\dot{\alpha}}} + i\theta^\alpha\partial_{\alpha\dot{\alpha}}, \quad \{D_\alpha, \bar{D}_{\dot{\alpha}}\} = 2i\partial_{\alpha\dot{\alpha}}. \quad \text{(A.20)}$$

A.4 The Grassmann integration conventions

$$\int d^2\theta\,\theta^2 = 1, \quad \int d^2\theta d^2\bar{\theta}\,\theta^2\bar{\theta}^2 = 1. \tag{A.21}$$

A.5 $(1,0)$ and $(1,0)$ sigma matrices

To convert the two-index spinorial symmetric representation in the vectorial representation we will need the following sigma matrices:

$$(\vec{\sigma})^{\alpha\beta} = \{\tau^3, i, -\tau^1\}_{\alpha\beta}, \quad (\vec{\sigma})_{\alpha\beta} = \{-\tau^3, i, \tau^1\}_{\alpha\beta},$$
$$(\vec{\sigma})^{\dot{\alpha}\dot{\beta}} = \{\tau^3, -i, -\tau^1\}_{\dot{\alpha}\dot{\beta}}, \quad (\vec{\sigma})_{\dot{\alpha}\dot{\beta}} = \{-\tau^3, -i, \tau^1\}_{\dot{\alpha}\dot{\beta}}. \tag{A.22}$$

A.6 The Weyl and Dirac spinors

If we have two Weyl (right-handed) spinors ξ^α and η_β, transforming in the representations R and $\bar{R}$ of the gauge group, respectively, then the Dirac spinor Ψ can be formed as

$$\Psi = \begin{pmatrix} \xi^\alpha \\ \bar{\eta}_{\dot{\alpha}} \end{pmatrix}. \tag{A.23}$$

The Dirac spinor Ψ has four components, while ξ^α and η_β have two components each.

A.7 Euclidean notation

As was mentioned, in Part II we switch to a formally Euclidean notation e.g.

$$F_{\mu\nu}^2 = 2F_{0i}^2 + F_{ij}^2, \tag{A.24}$$

and

$$(\partial_\mu a)^2 = (\partial_0 a)^2 + (\partial_i a)^2, \tag{A.25}$$

etc. This is appropriate, since we mostly consider static (time-independent) field configurations, and $A_0 = 0$. Then the Euclidean action is nothing but the energy functional.

Then, in the fermion sector we have to define the Euclidean matrices

$$(\sigma_\mu)^{\alpha\dot\alpha} = (1, -i\vec\tau)_{\alpha\dot\alpha}, \tag{A.26}$$

and

$$(\bar\sigma_\mu)_{\dot\alpha\alpha} = (1, i\vec\tau)_{\dot\alpha\alpha}. \tag{A.27}$$

Lowering and raising of the spinor indices is performed by virtue of the antisymmetric tensor defined as

$$\varepsilon_{12} = \varepsilon_{\dot1\dot2} = 1,$$
$$\varepsilon^{12} = \varepsilon^{\dot1\dot2} = -1. \tag{A.28}$$

The same raising and lowering convention applies to the flavor $SU(2)_R$ indices f, g, etc.

When the contraction of the spinor indices is assumed inside the parentheses we use the following notation:

$$(\lambda\psi) \equiv \lambda_\alpha\psi^\alpha, \quad (\bar\lambda\bar\psi) \equiv \bar\lambda^{\dot\alpha}\bar\psi_{\dot\alpha}. \tag{A.29}$$

A.8 Group-theory coefficients

As was mentioned, the gauge group is assumed to be $SU(N)$. For a given representation R of $SU(N)$, the definitions of the Casimir operators to be used below are

$$\mathrm{Tr}(T^a T^b)_R = T(R)\delta^{ab}, \quad (T^a T^a)_R = C(R)\,I, \tag{A.30}$$

where I is the unit matrix in this representation. It is quite obvious that

$$C(R) = T(R)\,\frac{\dim(G)}{\dim(R)}, \tag{A.31}$$

where $\dim(G)$ is the dimension of the group (= the dimension of the adjoint representation). Note that $T(R)$ is also known as (one half of) the Dynkin index, or the dual Coxeter number. For the adjoint representation, $T(R)$ is denoted by $T(G)$. Moreover, $T(SU(N)) = N$.

A.9 Renormalization-group conventions

We use the following definition of the β function (also known as the Gell–Mann–Low function) and anomalous dimensions:

$$\mu \frac{\partial \alpha}{\partial \mu} \equiv \beta(\alpha) = -\frac{\beta_0}{2\pi}\alpha^2 - \frac{\beta_1}{4\pi^2}\alpha^3 + \cdots \qquad (A.32)$$

while

$$\gamma = -d \ln Z(\mu)/d \ln \mu. \qquad (A.33)$$

In supersymmetric theories

$$\beta(\alpha) = -\frac{\alpha^2}{2\pi}\left[3\,T(G) - \sum_i T(R_i)(1 - \gamma_i)\right]\left(1 - \frac{T(G)\alpha}{2\pi}\right)^{-1}, \quad (A.34)$$

where the sum runs over all matter supermultiplets. This is the so-called Novikov–Shifman–Vainshtein–Zakharov (NSVZ) beta function [236]. The anomalous dimension of the ith matter superfield is

$$\gamma_i = -2C(R_i)\frac{\alpha}{2\pi} + \cdots \qquad (A.35)$$

Sometimes, when one-loop anomalous dimensions are discussed, the coefficient in front of $-\alpha/(2\pi)$ in (A.33) is also referred to as an "anomalous dimension."

Appendix B

Many faces of two-dimensional supersymmetric CP($N - 1$) model

B.1 O(3) sigma model

Supersymmetric extension of the O(3) sigma model in the form discussed in this section was suggested in Refs. [7, 237]. We refer the reader to the book [238] for a pedagogical discussion of the non-supersymmetric O(3) sigma model.

One can construct supersymmetric sigma model in terms of two-dimensional $\mathcal{N} = 1$ superfields as follows. Let us introduce a triplet of real superfields N^a,

$$N^a(x,\theta) = S^a(x) + \bar{\theta}\chi^a(x) + \frac{1}{2}\bar{\theta}\theta\, F^a(x), \quad a = 1, 2, 3, \tag{B.1}$$

where θ is a two-component Majorana (real) spinor ($\bar{\theta} = \theta\gamma^0$), χ^a is a two-component Majorana fermion field and F^a is an auxiliary boson field which will enter in the Lagrangian with no kinetic term. The superfield $N^a(x,\theta)$ is subject to the constraint

$$N^a(x,\theta)N^a(x,\theta) = 1. \tag{B.2}$$

In components this is equivalent to

$$S^a S^a = 1, \quad S^a\chi^a = 0, \quad S^a F^a = \frac{1}{2}\bar{\chi}^a\chi^a. \tag{B.3}$$

The action of the model takes the form

$$
\begin{aligned}
S &= \frac{1}{2g_0^2} \int d^2x\, d^2\theta\, \varepsilon^{\alpha\beta}\left(D_\alpha N^a\right)\left(D_\beta N^a\right) \\
&= \frac{1}{g_0^2} \int d^2x \left[\frac{1}{2}(\partial_\mu S^a)^2 + \frac{1}{2}\bar{\chi}^a i\gamma^\mu\partial_\mu\chi^a + \frac{1}{8}(\bar{\chi}\chi)^2\right]
\end{aligned}
\tag{B.4}
$$

where g_0^2 is the (bare) coupling constant and

$$D_\alpha = \frac{\partial}{\partial \bar{\theta}_\alpha} - i(\gamma^\mu \theta)_\alpha \, \partial_\mu. \tag{B.5}$$

This model describes two independent (real) degrees of freedom in the bosonic and fermionic sectors. The interaction inherent to this model is due to the constraints (B.3) and the four-fermion term in (B.4). The model is O(3) symmetric, by construction. Also by construction it has $\mathcal{N} = (1, 1)$ supersymmetry (i.e. one left-handed real supercharge, and one right-handed). In fact this model has an extended $\mathcal{N} = 2$ supersymmetry (more exactly, $\mathcal{N} = (2, 2)$). The occurrence of two extra supercharges (four altogether) is automatic and is explained by the fact that the target space of the bosonic sector is S^2, which is a Kähler manifold. Minimal $\mathcal{N} = (1, 1)$ supersymmetrization of any Kählerian sigma model automatically produces $\mathcal{N} = (2, 2)$ supersymmetry. Further details can be found in the review paper [156].

B.2 CP(1) sigma model

The same model expressed in terms of unconstrained variables is usually referred to as the CP(1) model. If the unit vector S^a parametrizes the sphere, one can pass to unconstrained variables by performing the stereographic projection of the sphere onto the complex ϕ plane,

$$\phi = \frac{S^1 + iS^2}{1 + S^3}. \tag{B.6}$$

The complex field ϕ replaces two independent components of S^a. The unconstrained two-component *complex* fermion field ψ is introduced as follows:

$$\psi = \frac{\chi^1 + i\chi^2}{1 + S^3} - \frac{S^1 + iS^2}{(1 + S^3)^2} \chi^3. \tag{B.7}$$

The inverse transformations have the form

$$S^1 = \frac{2(\mathrm{Re}\phi)}{1 + |\phi|^2}, \quad S^2 = \frac{2(\mathrm{Im}\phi)}{1 + |\phi|^2}, \quad S^3 = \frac{1 - |\phi|^2}{1 + |\phi|^2} \tag{B.8}$$

and

$$\chi^1 = \frac{2(\mathrm{Re}\,\psi)}{1 + |\phi|^2} - \frac{2(\mathrm{Re}\,\phi)[\phi^\dagger \psi + \mathrm{H.c.}]}{(1 + |\phi|^2)^2},$$

$$\chi^2 = \frac{2(\text{Im}\,\psi)}{1 + |\phi|^2} - \frac{2(\text{Im}\,\phi)[\phi^\dagger\psi + \text{H.c.}]}{(1 + |\phi|^2)^2},$$

$$\chi^3 = -2\frac{[\phi^\dagger\psi + \text{H.c.}]}{(1 + |\phi|^2)^2}. \tag{B.9}$$

Substituting Eqs. (B.8) and (B.9) in the action (B.4) we get [239]

$$L_{\text{CP}(1)} = G\left\{\partial_\mu\phi^\dagger\,\partial^\mu\phi + i\bar{\psi}\gamma^\mu\partial_\mu\psi - \frac{2i}{\chi}\,\phi^\dagger\partial_\mu\phi\,\bar{\psi}\gamma^\mu\psi + \frac{1}{\chi^2}\,(\bar{\psi}\psi)^2\right\} \tag{B.10}$$

where

$$G = \frac{2}{g_0^2\,\chi^2}, \quad \chi = 1 + |\phi|^2. \tag{B.11}$$

The above Lagrangian can be obtained in terms of $\mathcal{N} = 2$ superfields which will make its $\mathcal{N} = (2,2)$ supersymmetry explicit. Namely, let us introduce a chiral superfield

$$\Phi(x_L, \theta) = \phi(x_L) + \sqrt{2}\,\varepsilon_{\alpha\beta}\,\theta^\alpha\psi^\beta(x_L) + \varepsilon_{\alpha\beta}\,\theta^\alpha\theta^\beta\,F(x_L), \tag{B.12}$$

where θ is a two-component *complex* Grassmann variable, while

$$x_L^\mu = x^\mu + i\bar{\theta}\gamma^\mu\theta. \tag{B.13}$$

Moreover, $\Phi^\dagger$ depends on $x_R^\mu = x^\mu - i\bar{\theta}\gamma^\mu\theta$ and $\bar{\theta}$, a conjugation of (B.12). In terms of these superfields the Lagrangian of the CP(1) model can be written as

$$L_{\text{CP}(1)} = \int d^4\theta\, K(\Phi, \Phi^\dagger), \tag{B.14}$$

where K is the Kähler potential,

$$K = \frac{2}{g_0^2}\,\ln(1 + \Phi^\dagger\Phi). \tag{B.15}$$

Needless to say, $\mathcal{N} = 2$ supersymmetry is built in here. And what about the target space symmetry? The U(1) symmetry corresponding to the rotation around the third axis in the target space is realized linearly,

$$\Phi \to \Phi + i\alpha \cdot \Phi, \quad \Phi^\dagger \to \Phi^\dagger - i\alpha \cdot \Phi^\dagger, \tag{B.16}$$

where α is a real parameter. At the same time, two other symmetry rotations are realized nonlinearly,

$$\Phi \to \beta + \beta^* \cdot \Phi^2, \quad \Phi^\dagger \to \beta^* + \beta \cdot (\Phi^\dagger)^2, \tag{B.17}$$

with a complex parameter β.

B.3 Geometric interpretation

Equations (B.14) and (B.15) suggest a geometric interpretation (for a review see e.g. [240]) for the above formulation of the CP(1) model which, in turn, allows one to readily generalize it to the case of CP(N − 1) with arbitrary N. Indeed, let us consider $N - 1$ complex superfields

$$\Phi^i(x^\mu + i\bar{\theta}\gamma^\mu\theta), \quad \Phi^{\dagger \bar{j}}(x^\mu - i\bar{\theta}\gamma^\mu\theta),$$

and the Kähler potential

$$K = \frac{2}{g_0^2} \ln \left(1 + \sum_{i,\bar{j}=1}^{N-1} \Phi^{\dagger \bar{j}} \delta_{\bar{j}i} \Phi^i \right). \tag{B.18}$$

(As we will see momentarily, it corresponds to the so-called round Fubini–Study metric.) The Kähler potential determines the metric of the target space according to the formula

$$G_{i\bar{j}} = \frac{\partial^2 K(\phi, \phi^\dagger)}{\partial \phi^i \partial \phi^{\dagger \bar{j}}}. \tag{B.19}$$

For CP(N − 1) the Riemann tensor is expressed in terms of the metric (B.19) as follows:

$$R_{i\bar{j}k\bar{m}} = -\frac{g_0^2}{2} \left(G_{i\bar{j}} G_{k\bar{m}} + G_{i\bar{m}} G_{k\bar{j}}\right), \tag{B.20}$$

while the Ricci tensor

$$R_{i\bar{j}} = \frac{g_0^2}{2} N G_{i\bar{j}}. \tag{B.21}$$

In components the Lagrangian of the CP(N − 1) model takes the form [241]

$$L = \int d^4\theta \, K = G_{i\bar{j}} [\partial_\mu \phi^{\dagger \bar{j}} \, \partial_\mu \phi^i + i\bar{\psi}^{\bar{j}} \gamma^\mu D_\mu \psi^i] - \frac{1}{2} R_{i\bar{j}k\bar{l}} (\bar{\psi}^{\bar{j}} \psi^i)(\bar{\psi}^{\bar{l}} \psi^k), \tag{B.22}$$

where D is the covariant derivative,

$$D_\mu \psi^i = \partial_\mu \psi^i + \Gamma^i_{kl} (\partial_\mu \phi^k) \psi^l, \tag{B.23}$$

and Γ^i_{kl} is the Christoffel symbol.

If $N = 2$ the above expressions simplify and we get

$$G = G_{1\bar{1}} = \partial_\phi \partial_{\phi^\dagger} K \big|_{\theta=\bar{\theta}=0} = \frac{2}{g_0^2 \chi^2},$$

$$\Gamma = \Gamma_{11}^1 = -2\frac{\phi^\dagger}{\chi}, \quad \bar{\Gamma} = \Gamma_{\bar{1}\bar{1}}^{\bar{1}} = -2\frac{\phi}{\chi},$$

$$R \equiv R_{1\bar{1}} = -G^{-1}R_{1\bar{1}1\bar{1}} = \frac{2}{\chi^2}, \tag{B.24}$$

where we use the notation

$$\chi \equiv 1 + \phi\phi^\dagger. \tag{B.25}$$

Substituting (B.24) and (B.25) in (B.22) we arrive at the CP(1) Lagrangian (B.10).

B.4 Gauged formulation

Here we will discuss yet another formulation of $\mathcal{N} = 2$ supersymmetric sigma models with the target space

$$\frac{\text{SU}(N)}{\text{SU}(N-1) \times \text{U}(1)} = \text{CP}(N-1), \tag{B.26}$$

which goes under the name of the gauged formulation [242]. This formulation is built on an N-plet of complex scalar fields n^i where $i = 1, 2, ..., N$. We impose the constraint

$$n_i^\dagger n^i = 1. \tag{B.27}$$

This leaves us with $2N - 1$ real bosonic degrees of freedom. To eliminate one extra degree of freedom we impose a local U(1) invariance $n^i(x) \to e^{i\alpha(x)}n^i(x)$. To this end we introduce a gauge field A_μ which converts the partial derivative into the covariant one,

$$\partial_\mu \to \nabla_\mu \equiv \partial_\mu - i A_\mu. \tag{B.28}$$

The field A_μ is auxiliary; it enters in the Lagrangian without derivatives. The kinetic term of the n fields is

$$L = \frac{2}{g_0^2} \left| \nabla_\mu n^i \right|^2. \tag{B.29}$$

The superpartner to the field n^i is an N-plet of complex two-component spinor fields ξ^i,

$$\xi^i = \begin{cases} \xi_R^i \\ \xi_L^i \end{cases}.$$ (B.30)

The auxiliary field A_μ has a complex scalar superpartner σ and a two-component complex spinor superpartner λ; both enter without derivatives. The full $\mathcal{N} = 2$ symmetric Lagrangian is

$$L = \frac{2}{g_0^2}\Big\{ |\nabla_\mu n^i|^2 + \bar{\xi}_i\, i\gamma^\mu \nabla_\mu \xi^i + 2|\sigma|^2\, |n^i|^2$$

$$+ \Big[i\sqrt{2}\,\sigma\xi_{iR}^\dagger \xi_L^i + i\sqrt{2}\, n_i^\dagger (\lambda_R \xi_L^i - \lambda_L \xi_R^i) + \text{H.c.}\Big]\Big\}. \quad (B.31)$$

The auxiliary fields can be eliminated by virtue of the equations of motion which yield the following relations:

$$n_i^\dagger \xi_L^l = 0, \quad n_i^\dagger \xi_R^l = 0;$$

$$A_\mu = -\frac{i}{2}\, n_l^\dagger \overset{\leftrightarrow}{\partial}_\mu n^l - \frac{1}{2}\bar{\xi}_l\gamma_\mu\xi^l,$$

$$\sigma = \frac{i}{\sqrt{2}}\, \xi_{iL}^\dagger \xi_R^l. \quad (B.32)$$

Substituting (B.32) in (B.31) we arrive at the final expression for the Lagrangian of $\mathcal{N} = 2$ sigma model with the target space (B.26),

$$L = \frac{2}{g_0^2}\Big\{ |\partial_\mu n^i|^2 + \frac{1}{4}(n_i^\dagger \overset{\leftrightarrow}{\partial}_\mu n^i)^2$$

$$+ \bar{\xi}_i\, i\gamma^\mu \Big(\partial_\mu - \frac{1}{2}n_l^\dagger \overset{\leftrightarrow}{\partial}_\mu n^l \Big)\xi^i$$

$$- \big(\xi_{iR}^\dagger \xi_R^i \cdot \xi_{iL}^\dagger \xi_L^l + \xi_{iR}^\dagger \xi_L^i \cdot \xi_{iL}^\dagger \xi_R^l \big)\Big\}, \quad (B.33)$$

$$n_i^\dagger n^i = 1, \quad n_i^\dagger \xi^i = 0. \quad (B.34)$$

For $N = 2$ there exists a simple local transformation converting the Lagrangian of the O(3) model discussed in Appendix B.1 into (B.33),

$$S^a = n_i^\dagger (\tau^a)_k^i\, n^k,$$

$$\chi^a = n_i^\dagger (\tau^a)_k^i\, \xi^k + \xi_i^\dagger (\tau^a)_k^i\, n^k, \quad (B.35)$$

where τ^a are the Pauli matrices. If we use the Fierz identity for the Pauli matrices,

$$(\tau^a)^i_k \, (\tau^a)^{\tilde{i}}_{\tilde{k}} = -\frac{1}{2}(\tau^a)^{\tilde{i}}_k (\tau^a)^i_{\tilde{k}} + \frac{3}{2}\delta^{\tilde{i}}_k \delta^i_{\tilde{k}}, \tag{B.36}$$

and substitute Eq. (B.35) in the Lagrangian (B.4) taking account of the constraints (B.3) we arrive at (B.33). The constraints (B.34) are satisfied automatically.

B.5 Heterotic CP(1)

Here we will outline derivation of the heterotic CP(1) model elaborated in Ref. [191]. We will start from the general geometric formulation presented in Appendix B.3, specify it to the CP(1) case using Eq. (B.24) and then introduce a deformation that breaks $\mathcal{N} = (2,2)$ down to $\mathcal{N} = (0,2)$. As is well known, if we limit ourselves to the set of fields present in the $\mathcal{N} = (2,2)$ sigma model, such a deformation does not exist. However, it does exist if we agree to introduce an extra *right-handed* fermion ζ_R [190].

One can obtain the deformed Lagrangian as follows. Introduce the operators

$$\begin{aligned}
\mathcal{B} &= \{\zeta_R(x^\mu + i\bar{\theta}\gamma^\mu\theta) + \sqrt{2}\theta_R\mathcal{F}\}\theta^\dagger_L, \\
\mathcal{B}^\dagger &= \theta_L\{\zeta^\dagger_R(x^\mu - i\bar{\theta}\gamma^\mu\theta) + \sqrt{2}\theta^\dagger_R\mathcal{F}^\dagger\}.
\end{aligned} \tag{B.37}$$

Since θ_L and $\theta^\dagger_L$ enter in Eq. (B.37) explicitly, $\mathcal{B}$ and $\mathcal{B}^\dagger$ are *not* superfields with regards to the supertransformations with parameters ϵ_L, $\epsilon^\dagger_L$. These supertransformations are absent in the heterotic model. Only those survive which are associated with ϵ_R, $\epsilon^\dagger_R$. Note that $\mathcal{B}$ and $\mathcal{B}^\dagger$ are superfields with regards to the latter.

It is convenient to introduce a shorthand for the chiral coordinate

$$\tilde{x}^\mu = x^\mu + i\bar{\theta}\gamma^\mu\theta. \tag{B.38}$$

Then the transformation laws with the parameters ϵ_R, $\epsilon^\dagger_R$ are as follows:

$$\delta\theta_R = \epsilon_R, \quad \delta\theta^\dagger_R = \epsilon^\dagger_R, \quad \delta\tilde{x}^0 = 2i\epsilon^\dagger_R\theta_R, \quad \delta\tilde{x}^1 = 2i\epsilon^\dagger_R\theta_R. \tag{B.39}$$

With respect to such supertransformations, $\mathcal{B}$ and $\mathcal{B}^\dagger$ are superfields. Indeed,

$$\delta\zeta_R = \sqrt{2}\,\mathcal{F}\epsilon_R, \quad \delta\mathcal{F} = \sqrt{2}\,i(\partial_L\zeta_R)\epsilon^\dagger_R, \tag{B.40}$$

plus Hermitean conjugate transformations. To convert $L_{CP(1)}$ into $L_{heterotic}$ we add to $L_{CP(1)}$ the following terms:

$$\Delta L = \int d^4\theta\{-2\mathcal{B}^\dagger\mathcal{B} + [g^2_0\sqrt{2}\,\gamma\,\mathcal{B}\,K + \text{H.c.}]\}, \tag{B.41}$$

where γ is generally speaking a complex constant. For simplicity we will assume γ to be real. Thus, we obviously deal here with a single deformation parameter.

First, let us check that the extra term (B.41) preserves invariance on the target space. Indeed, the invariance under the U(1) transformation of the superfields Φ, $\Phi^\dagger$,

$$\Phi \to i\delta \Phi, \quad \Phi^\dagger \to -i\delta \Phi^\dagger, \tag{B.42}$$

is obvious. Two other rotations on the sphere manifest themselves in nonlinear transformations with a complex parameter β,

$$\Phi \to \beta + \beta^* \Phi^2, \quad \Phi^\dagger \to \beta^* + \beta(\Phi^\dagger)^2. \tag{B.43}$$

Under these transformations

$$\delta K = \frac{2}{g_0^2} (\beta^* \Phi + \beta \Phi^\dagger). \tag{B.44}$$

It is not difficult to see that

$$\int d^4\theta \, B \, \delta K = 0. \tag{B.45}$$

In other words, even before performing the component decomposition we are certain that the term (B.41) is invariant on the target space of the CP(1) model. Needless to say, it is $\mathcal{N} = (0, 2)$ invariant by construction.

As usual, the $\mathcal{F}$ term enters without derivatives and can be eliminated by virtue of equations of motion,

$$\mathcal{F} = -2\gamma^* \chi^{-2} \psi_R^\dagger \psi_L, \quad \mathcal{F}^\dagger = -2\gamma \chi^{-2} \psi_L^\dagger \psi_R. \tag{B.46}$$

In addition, the F terms of the superfields Φ, $\Phi^\dagger$ also change. If before the deformation e.g. $F = (i/2) \Gamma \psi \gamma^0 \psi$, after the deformation

$$F = \frac{i}{2} \Gamma \psi \gamma^0 \psi - g_0^2 \gamma \psi_L \zeta_R^\dagger, \tag{B.47}$$

plus the Hermitian conjugated expression for $F^\dagger$.

Assembling all these pieces together we get the Lagrangian of the heterotic CP(1) model,

$$
\begin{aligned}
L_{\text{heterotic}} = {}& \zeta_R^\dagger \, i\partial_L \, \zeta_R + \left[\gamma \, \zeta_R \, R \left(i \, \partial_L \phi^\dagger \right) \psi_R + \text{H.c.} \right] - g_0^2 |\gamma|^2 (\zeta_R^\dagger \, \zeta_R)(R \, \psi_L^\dagger \psi_L) \\
& + G \left\{ \partial_\mu \phi^\dagger \, \partial^\mu \phi + \frac{i}{2} (\psi_L^\dagger \overset{\leftrightarrow}{\partial}_R \psi_L + \psi_R^\dagger \overset{\leftrightarrow}{\partial}_L \psi_R) \right. \\
& \qquad - \frac{i}{\chi} \left[\psi_L^\dagger \psi_L (\phi^\dagger \overset{\leftrightarrow}{\partial}_R \phi) + \psi_R^\dagger \psi_R (\phi^\dagger \overset{\leftrightarrow}{\partial}_L \phi) \right] \\
& \qquad \left. - \frac{2(1 - g_0^2 |\gamma|^2)}{\chi^2} \, \psi_L^\dagger \psi_L \psi_R^\dagger \psi_R \right\},
\end{aligned}
\tag{B.48}
$$

where R stands for the Ricci tensor, and

$$
\partial_L = \frac{\partial}{\partial t} + \frac{\partial}{\partial z}, \qquad \partial_R = \frac{\partial}{\partial t} - \frac{\partial}{\partial z}.
\tag{B.49}
$$

Generalization for arbitrary N (i.e. the $\mathcal{N} = (0,2)$ deformed CP($N-1$) model) is as follows:

$$
\begin{aligned}
L_{\text{heterotic}} = {}& \zeta_R^\dagger \, i\partial_L \, \zeta_R + \left[\gamma \, g_0^2 \, \zeta_R \, G_{i\bar{j}} \left(i \, \partial_L \phi^{\dagger \bar{j}} \right) \psi_R^i + \text{H.c.} \right] \\
& - g_0^4 |\gamma|^2 (\zeta_R^\dagger \, \zeta_R)(G_{i\bar{j}} \, \psi_L^{\dagger \bar{j}} \psi_L^i) \\
& + G_{i\bar{j}} [\partial_\mu \phi^{\dagger \bar{j}} \, \partial_\mu \phi^i + i \bar{\psi}^{\bar{j}} \gamma^\mu D_\mu \psi^i] \\
& - \frac{g_0^2}{2} (G_{i\bar{j}} \psi_R^{\dagger \bar{j}} \, \psi_R^i)(G_{k\bar{m}} \psi_L^{\dagger \bar{m}} \, \psi_L^k) \\
& + \frac{g_0^2}{2} (1 - 2g_0^2 |\gamma|^2)(G_{i\bar{j}} \psi_R^{\dagger \bar{j}} \, \psi_L^i)(G_{k\bar{m}} \psi_L^{\dagger \bar{m}} \, \psi_R^k).
\end{aligned}
\tag{B.50}
$$

Appendix C

Strings in $\mathcal{N} = 2$ SQED

In this Appendix we briefly review the Abelian Abrikosov–Nielsen–Olesen strings in $\mathcal{N} = 2$ supersymmetric QED in four dimensions. The BPS strings in this theory were first considered in [148, 35].

C.1 $\mathcal{N} = 2$ supersymmetric QED

$\mathcal{N} = 2$ supersymmetric QED is discussed in Section 8.1. Here we summarize basic features of this theory for convenience. The field content of $\mathcal{N} = 2$ supersymmetric QED consists of a U(1) vector $\mathcal{N} = 2$ multiplet as well as N_f matter hypermultiplets. The mass terms are introduced via the superpotential

$$W = \sum_A \left(m_A Q^A \tilde{Q}_A + \frac{1}{\sqrt{2}} A Q^A \tilde{Q}_A \right). \tag{C.1}$$

For the definition of the Fayet–Iliopoulos term see Eq. (3.2.1). In this form it is the same in $\mathcal{N} = 1$ and $\mathcal{N} = 2$, cf. Eq. (4.1.5). The bosonic part of the action of this theory is

$$S = \int d^4x \left\{ \frac{1}{4g^2} F_{\mu\nu}^2 + \frac{1}{g^2} |\partial_\mu a|^2 + \bar{\nabla}_\mu \bar{q}_A \nabla_\mu q^A + \bar{\nabla}_\mu \tilde{q}_A \nabla_\mu \bar{\tilde{q}}^A \right.$$
$$+ n_e^2 \frac{g^2}{2} (|q^A|^2 - |\tilde{q}_A|^2 - \xi)^2 + 2n_e^2 g^2 |\tilde{q}_A q^A|^2$$
$$\left. + \frac{1}{2} (|q^A|^2 + |\tilde{q}^A|^2) |a + \sqrt{2} m_A|^2 \right\}, \tag{C.2}$$

where

$$\nabla_\mu = \partial_\mu - i n_e A_\mu, \quad \bar{\nabla}_\mu = \partial_\mu + i n_e A_\mu. \tag{C.3}$$

243

Here ξ is the coefficient in front of the Fayet–Iliopoulos term; we consider the FI D-term here while g is the U(1) gauge coupling and n_e is the electric charge. It can be integer or half integer. The index $A = 1, \ldots, N_f$ is the flavor index. Below we consider the case $N_f = 1$. This is the simplest case which admits BPS string solutions.

The FI term triggers the squark condensation. The vacuum of this theory is given by

$$a = -\sqrt{2}m, \quad q = \sqrt{\xi}, \quad \tilde{q} = 0. \tag{C.4}$$

Hereafter in search for string solutions we will stick to the *ansatz* $\tilde{q} = 0$.

Now let us discuss the mass spectrum of the light fields in this vacuum. The spectrum can be obtained by diagonalizing the quadratic form in (C.2). This is done in Ref. [35]; the result is as follows: one real component of the field q is eaten up by the Higgs mechanism to become the third component of the massive photon. Three components of the massive photon, one remaining component of q and four real components of the fields $\tilde{q}$ and a form one long $\mathcal{N} = 2$ multiplet (8 boson states + 8 fermion states), with mass

$$m_\gamma^2 = 2n_e^2 g^2 \xi. \tag{C.5}$$

C.2 String solutions

As soon as fields a and $\tilde{q}$ play no role in string solutions we can look for these solutions using the reduced theory with these fields set to zero. The bosonic action (C.2) reduces to

$$S = \int d^4x \left\{ \frac{1}{4g^2} F_{\mu\nu}^2 + |\nabla_\mu q|^2 + \frac{g^2}{2} n_e^2 \left(|q|^2 - \xi \right)^2 \right\}. \tag{C.6}$$

Since the U(1) gauge group is spontaneously broken, the model supports conventional ANO strings [36]. The topological stability of the ANO string is due to the fact that $\pi_1(U(1)) = Z$.

Let us derive the first-order equations which determine the string solution making use of the Bogomol'nyi representation [5] of the model (C.6). We have for the string tension

$$T = \int d^2x \left\{ \left[\frac{1}{\sqrt{2}g} F_3^* + \frac{g}{\sqrt{2}} n_e \left(|q|^2 - \xi \right) \right]^2 \right.$$

$$\left. + \left| \nabla_1 q + i\nabla_2 q \right|^2 + n_e \xi F_3^* \right\}, \tag{C.7}$$

where $F_3^* = F_{12}$ and we assume that the fields in this expression depend only on the coordinates $x_i, i = 1, 2$.

The Bogomol'nyi representation (C.7) leads us to the following first-order equations:

$$F_3^* + g n_e \left(|q|^2 - \xi \right) = 0,$$
$$(\nabla_1 + i \nabla_2) q = 0. \tag{C.8}$$

Once these equations are satisfied the energy of the BPS object is given by the last surface term in (C.7). Note that representation (C.7) can be written also with different sign in front of the terms proportional to the gauge fluxes. This would give first-order equations for the anti-string, with negative values of gauge fluxes.

For the topologically stable string solution, the scalar field winds n times in U(1) gauge group when we move around the string along a large circle in the (x, y) plane (we assume that the string stretches along the z-axis),

$$q \sim \sqrt{\xi} \, e^{i n \alpha},$$
$$A_i \sim \frac{n}{n_e} \partial_i \alpha, \quad r \to \infty, \tag{C.9}$$

where r and α are the polar coordinates in the (x, y) plane (see Fig. 3.6) and $i = 1, 2$. This ensures that the flux of the string is

$$\int d^2 x \, F_3^* = \frac{2 \pi n}{n_e}. \tag{C.10}$$

The tension of the string with winding number n is determined by the surface term in (C.7),

$$T_n = 2 \pi n \, \xi. \tag{C.11}$$

For the elementary $n = 1$ string the solution can be found using the standard *ansatz* [5]

$$q(x) = \phi(r) \, e^{i \alpha}, \quad A_i(x) = \frac{1}{n_e} \partial_i \alpha \left[1 - f(r) \right], \tag{C.12}$$

where we introduced two profile functions ϕ and f for the scalar and gauge fields, respectively.

The *ansatz* (C.12) goes through the set of equations (C.8), and we get the following two equations for the profile functions:

$$-\frac{1}{r} \frac{df}{dr} + n_e^2 g^2 (\phi^2 - \xi) = 0, \quad r \frac{d\phi}{dr} - f \phi = 0. \tag{C.13}$$

The boundary conditions for the profile functions are the following. At large distances we have

$$\phi(\infty) = \sqrt{\xi}, \quad f(\infty) = 0. \tag{C.14}$$

At the origin the smoothness of the field configuration at hand requires

$$\phi(0) = 0, \quad f(0) = 1. \tag{C.15}$$

These boundary conditions are such that the scalar field reaches its vacuum value at infinity. The same first-order equations arise for the BPS vortex in $\mathcal{N} = 1$ QED in $(2+1)$ dimensions, see Chapter 3. The fermion zero modes for the BPS vortices in $(3+1)$ and $(2+1)$ dimensions are different, however. Equations (C.13) have a unique solution for the profile functions, which can be found numerically [4], see Fig. 3.7. The string transverse size is $\sim 1/m_\gamma$.

First-order equations (C.13) can also be obtained using supersymmetry. We start from the supersymmetry transformations for the fermion fields in the theory (C.2),

$$\delta\lambda^{\alpha f} = \frac{1}{2}(\sigma_\mu \bar{\sigma}_\nu \epsilon^{\,f})^\alpha F_{\mu\nu} + \epsilon^{\alpha p} F^m (\tau^m)_p^{\,f} + \dots,$$

$$\delta\bar{\tilde{\psi}}_{\dot\alpha}^A = i\sqrt{2}\,\bar{\mathcal{Y}}_{\dot\alpha\alpha} q_f^A \epsilon^{\alpha f} + \dots,$$

$$\delta\bar{\psi}_{\dot\alpha A} = i\sqrt{2}\,\bar{\mathcal{Y}}_{\dot\alpha\alpha} \bar{q}_{fA} \epsilon^{\alpha f} + \dots \tag{C.16}$$

Here $f = 1, 2$ is the $SU(2)_R$ index so $\lambda^{\alpha f}$ are the fermions from the $\mathcal{N} = 2$ vector supermultiplet, while q^{Af} denotes the $SU(2)_R$ doublet of the squark fields q^A and $\bar{\tilde{q}}^A$ in the quark hypermultiplets. The parameters of the SUSY transformations are denoted as $\epsilon^{\alpha f}$. Furthermore, the F terms in Eq. (C.16) are

$$F^3 = -i\, n_e\, g^2\, (\mathrm{Tr}\, |q|^2 - \xi), \quad F^1 + i F^2 = 0. \tag{C.17}$$

The dots in (C.16) stand for terms involving the a field which vanish on the string solution because it is given by its vacuum expectation value (C.4).

In Ref. [35] it was shown that four (real) supercharges generated by

$$\epsilon^{12}, \quad \epsilon^{21} \tag{C.18}$$

act trivially on the BPS string. Namely imposing conditions $\epsilon^{11} = \epsilon^{22} = 0$ and requiring that the left-hand sides of Eqs. (C.16) are zero we get the first-order equations (C.13) upon substitution of the *ansatz* (C.12).[1]

---

[1] If we instead of (C.18) impose that different combinations of SUSY transformation parameters act trivially we get the equations for anti-string with the opposite directions of gauge fluxes.

C.3 The fermion zero modes

The string is half-critical, so 1/2 of the supercharges (related to the SUSY transformation parameters ϵ^{12} and ϵ^{21}), act trivially on the string solution. The remaining four (real) supercharges parametrized by ϵ^{11} and ϵ^{22} generate four (real) supertranslational fermion zero modes. They have the form [35]

$$\bar{\psi}_{\dot{2}} = -2\sqrt{2}\,\frac{x_1 + ix_2}{r^2}\, f\,\phi\,\alpha^{11},$$
$$\bar{\psi}_{\dot{1}} = 2\sqrt{2}\,\frac{x_1 - ix_2}{r^2}\, f\,\phi\,\alpha^{22},$$
$$\bar{\psi}_{\dot{1}} = 0, \quad \bar{\psi}_{\dot{2}} = 0,$$
$$\lambda^{11} = -in_e g^2(\phi^2 - \xi)\alpha^{11},$$
$$\lambda^{22} = in_e g^2(\phi^2 - \xi)\alpha^{22},$$
$$\lambda^{12} = 0, \quad \lambda^{21} = 0, \tag{C.19}$$

where the modes proportional to complex Grassmann parameters α^{11} and α^{22} are generated by ϵ^{11} and ϵ^{22} transformations, respectively.

References

[1] E. Witten and D. I. Olive, *Phys. Lett. B* **78**, 97 (1978).
[2] N. Seiberg and E. Witten, *Nucl. Phys. B* **426**, 19 (1994), (E) *B* 430, 485 (1994) [hep-th/9407087].
[3] N. Seiberg and E. Witten, *Nucl. Phys. B* 431, 484 (1994) [hep-th/9408099].
[4] H. J. de Vega and F. A. Schaposnik, *Phys. Rev. D* **14**, 1100 (1976), reprinted in *Solitons and Particles*, eds. C. Rebbi and G. Soliani (Singapore: World Scientific, 1984) p. 382.
[5] E. B. Bogomol'nyi, *Yad. Fiz.* **24**, 861 (1976) [*Sov. J. Nucl. Phys.* **24**, 449 (1976)], reprinted in *Solitons and Particles*, eds. C. Rebbi and G. Soliani (Singapore: World Scientific, 1984) p. 389.
[6] M. K. Prasad and C. M. Sommerfield, *Phys. Rev. Lett.* **35**, 760 (1975), reprinted in *Solitons and Particles*, eds. C. Rebbi and G. Soliani (Singapore: World Scientific, 1984) p. 530.
[7] P. Di Vecchia and S. Ferrara, *Nucl. Phys. B* **130**, 93 (1977).
[8] A. D'Adda, R. Horsley, and P. Di Vecchia, *Phys. Lett. B* **76**, 298 (1978).
[9] J. Hruby, *Nucl. Phys. B* **162**, 449 (1980).
[10] J. Polchinski, *Phys. Rev. Lett.* **75**, 4724 (1995) [hep-th/9510017]; see also the excellent text by J. Polchinski, *String Theory*, Vols. 1 and 2 (Cambridge: Cambridge University Press, 1998).
[11] G. R. Dvali and M. A. Shifman, *Phys. Lett. B* **396**, 64 (1997) (E) *B* **407**, 452 (1997) [hep-th/9612128].
[12] E. Witten, *Nucl. Phys. B* **507**, 658 (1997) [hep-th/9706109].
[13] D. Tong, *TASI Lectures on Solitons,* hep-th/0509216.
[14] M. Eto, Y. Isozumi, M. Nitta, K. Ohashi, and N. Sakai, *J. Phys. A* **39**, R315 (2006) [hep-th/0602170].
[15] F. A. Schaposnik, *Vortices,* hep-th/0611028.
[16] J. A. Harvey, Magnetic monopoles, duality, and supersymmetry, in *Fields, Strings and Duality: TASI 96*, eds. C. Efthimiou and B. Greene (Singapore: World Scientific, 1997), pp. 157–216 [hep-th/9603086].
[17] K. Konishi, *The magnetic monopoles seventy five years later*, hep-th/0702102.
[18] Y. A. Golfand and E. P. Likhtman, Pisma Zh. Eksp. Teor. Fiz. **13** (1971) 452 [JETP Lett. **13**, 323 (1971); reprinted in *Supersymmetry*, ed. S. Ferrara (North-Holland/World Scientific, 1987) Vol. 1, p. 7].
[19] J. T. Lopuszanski and M. Sohnius, Karlsruhe Report Print-74-1269 (unpublished).

[20] R. Haag, J. T. Lopuszanski, and M. Sohnius, *Nucl. Phys. B* **88**, 257 (1975) [Reprinted in *Supersymmetry*, ed. S. Ferrara (North-Holland/World Scientific, 1987) Vol. 1, p. 51].

[21] S. Gates, Jr., M. Grisaru, M. Roček, and W. Siegel, *Superspace* (Reading, MA: Benjamin/Cummings, 1983).

[22] J. W. van Holten and A. Van Proeyen, *J. Phys. A* **15**, 3763 (1982).

[23] J. A. de Azcarraga, J. P. Gauntlett, J. M. Izquierdo, and P. K. Townsend, *Phys. Rev. Lett.* **63**, 2443 (1989).

[24] E. R. Abraham and P. K. Townsend, *Nucl. Phys. B* **351**, 313 (1991).

[25] P. K. Townsend, P-brane democracy, in *The World in Eleven Dimensions: Supergravity, Supermembranes and M-theory*, ed. M. Duff (Bristol: IOP, 1999) pp. 375–89 [hep-th/9507048].

[26] S. Ferrara and M. Porrati, *Phys. Lett. B* **423**, 255 (1998) [hep-th/9711116].

[27] A. Gorsky and M. Shifman, *Phys. Rev. D* **61**, 085001 (2000) [hep-th/9909015].

[28] Z. Hloušek and D. Spector, *Nucl. Phys. B* **370**, 143 (1992); J. D. Edelstein, C. Nuñez, and F. Schaposnik, *Phys. Lett. B* **329**, 39 (1994) [hep-th/9311055]; S. C. Davis, A. C. Davis, and M. Trodden, *Phys. Lett. B* **405**, 257 (1997) [hep-ph/9702360].

[29] M. A. Shifman and A. I. Vainshtein, Instantons versus supersymmetry: fifteen years later, in M. Shifman, *ITEP Lectures in Particle Physics and Field Theory* (Singapore: World Scientific, 1999), Vol. 2, pp. 485–648 [hep-th/9902018].

[30] N. Dorey, JHEP **9811**, 005 (1998) [hep-th/9806056].

[31] A. Hanany and K. Hori, *Nucl. Phys. B* **513**, 119 (1998) [hep-th/9707192].

[32] L. Alvarez-Gaumé and D. Z. Freedman, *Commun. Math. Phys.* **91**, 87 (1983); S. J. Gates, *Nucl. Phys. B* **238**, 349 (1984); S. J. Gates, C. M. Hull, and M. Roček, *Nucl. Phys. B* **248**, 157 (1984).

[33] A. Losev and M. Shifman, *Phys. Rev. D* **68**, 045006 (2003) [hep-th/0304003].

[34] M. Shifman, A. Vainshtein, and R. Zwicky, *J. Phys. A* **39**, 13005 (2006) [hep-th/0602004].

[35] A. I. Vainshtein and A. Yung, *Nucl. Phys. B* **614**, 3 (2001) [hep-th/0012250].

[36] A. Abrikosov, *Sov. Phys. JETP* **32** 1442 (1957) [Reprinted in *Solitons and Particles*, eds. C. Rebbi and G. Soliani (Singapore: World Scientific, 1984), p. 356]; H. Nielsen and P. Olesen, *Nucl. Phys. B* **61**, 45 (1973) [Reprinted in *Solitons and Particles*, eds. C. Rebbi and G. Soliani (Singapore: World Scientific, 1984), p. 365].

[37] M. Shifman and A. Yung, *Phys. Rev. D* **70**, 025013 (2004) [hep-th/0312257].

[38] A. Rebhan, P. van Nieuwenhuizen and R. Wimmer, *Phys. Lett. B* **594**, 234 (2004) [hep-th/0401116].

[39] M. Shifman and A. Yung, *Phys. Rev. D* **70**, 045004 (2004) [hep-th/0403149].

[40] J. Wess and J. Bagger, *Supersymmetry and Supergravity*, Second Edition (Princeton University Press, 1992).

[41] A. D'Adda and P. Di Vecchia, *Phys. Lett. B* **73**, 162 (1978).

[42] R. K. Kaul and R. Rajaraman, *Phys. Lett. B* **131**, 357 (1983).

[43] C. Imbimbo and S. Mukhi, *Nucl. Phys. B* **247**, 471 (1984).

[44] A. D'Adda, R. Horsley, and P. Di Vecchia, *Phys. Lett. B* **76**, 298 (1978); R. Horsley, *Nucl. Phys. B* **151**, 399 (1979).

[45] J. F. Schönfeld, *Nucl. Phys. B* **161** (1979) 125.

[46] S. Rouhani, *Nucl. Phys. B* **182**, 462 (1981).

[47] A. Uchiyama, *Nucl. Phys. B* **244** (1984) 57; *Prog. Theor. Phys.* **75**, 1214 (1986).

[48] A. Uchiyama, *Nucl. Phys. B* **278**, 121 (1986).

[49] H. Yamagishi, *Phys. Lett. B* **147**, 425 (1984).

[50] A. K. Chatterjee and P. Majumdar, *Phys. Lett. B* **159**, 37 (1985).

[51] A. Chatterjee and P. Majumdar, *Phys. Rev. D* **30**, 844 (1984).

[52] A. Rebhan and P. van Nieuwenhuizen, *Nucl. Phys. B* **508**, 449 (1997) [hep-th/9707163].

[53] H. Nastase, M. A. Stephanov, P. van Nieuwenhuizen, and A. Rebhan, *Nucl. Phys. B* **542**, 471 (1999) [hep-th/9802074].

[54] N. Graham and R. L. Jaffe, *Nucl. Phys. B* **544**, 432 (1999) [hep-th/9808140].

[55] M. A. Shifman, A. I. Vainshtein, and M. B. Voloshin, *Phys. Rev. D* **59**, 045016 (1999) [hep-th/9810068].

[56] A. Losev, M. A. Shifman, and A. I. Vainshtein, *New J. Phys.* **4**, 21 (2002) [hep-th/0011027]; *Phys. Lett. B* **522**, 327 (2001) [hep-th/0108153].

[57] A. S. Goldhaber, A. Rebhan, P. van Nieuwenhuizen, and R. Wimmer, *Phys. Reports* **398**, 179 (2004) [hep-th/0401152].

[58] A. S. Goldhaber, A. Litvintsev, and P. van Nieuwenhuizen, *Phys. Rev. D* **67**, 105021 (2003) [arXiv:hep-th/0109110]; A. Rebhan, P. van Nieuwenhuizen, and R. Wimmer, *Nucl. Phys. B* **648**, 174 (2003) [arXiv:hep-th/0207051].

[59] P. Fendley, S. D. Mathur, C. Vafa, and N. P. Warner, *Phys. Lett. B* **243**, 257 (1990).

[60] M. Cvetic, F. Quevedo, and S. J. Rey, *Phys. Rev. Lett.* **67**, 1836 (1991).

[61] S. Cecotti and C. Vafa, *Commun. Math. Phys.* **158**, 569 (1993) [hep-th/9211097].

[62] B. Chibisov and M. A. Shifman, *Phys. Rev. D* **56**, 7990 (1997), (E) *D* **58**, 109901 (1998), [hep-th/9706141].

[63] D. Bazeia, J. Menezes, and M. M. Santos, *Nucl. Phys. B* **636**, 132 (2002) [hep-th/0103041]; *Phys. Lett. B* **521**, 418 (2001) [hep-th/0110111].

[64] P. K. Townsend, *Phys. Lett. B* **202**, 53 (1988).

[65] J. Wess and B. Zumino, *Phys. Lett. B* **49** (1974) 52.

[66] J. Iliopoulos and B. Zumino, *Nucl. Phys. B* **76** (1974) 310; P. West, *Nucl. Phys. B* **106** (1976) 219; M. Grisaru, M. Roček, and W. Siegel, *Nucl. Phys. B* **159** (1979) 429.

[67] B. S. Acharya and C. Vafa, *On domain walls of* $\mathcal{N} = 1$ *supersymmetric Yang–Mills in four dimensions,* hep-th/0103011.

[68] I. I. Kogan, M. A. Shifman, and A. I. Vainshtein, *Phys. Rev. D* **53**, 4526 (1996) [Erratum-ibid. D **59**, 109903 (1999)] [arXiv:hep-th/9507170].

[69] T. E. Clark, O. Piguet, and K. Sibold, *Nucl. Phys. B* **159**, 1 (1979); K. Konishi, *Phys. Lett. B* **135**, 439 (1984); K. Konishi and K. Shizuya, *Nuovo Cim. A* **90**, 111 (1985).

[70] M. A. Shifman and A. I. Vainshtein, *Nucl. Phys. B* **296**, 445 (1988).

[71] N. Seiberg, The power of holomorphy: Exact results in 4D SUSY field theories, in *Proc. VI International Symposium on Particles, Strings, and Cosmology (PASCOS 94),* ed. K. C. Wali (Singapore: World Scientific, 1995) [hep-th/9408013]; K. Intriligator and N. Seiberg, *Lectures on Supersymmetric Gauge Theories and Electric–Magnetic Duality, Nucl. Phys. Proc. Suppl.* **45BC** (1996) 1 [hep-th/9509066]; K. Intriligator and N. Seiberg, Phases of $\mathcal{N} = 1$ supersymmetric gauge theories and electric–magnetic triality, in *Proc. Conf. Future Perspectives in String Theory (Strings '95)* eds. I. Bars, P. Bouwknegt, J. Minahan, D. Nemeschansky, K. Pilch, H. Saleur, and N. Warner (Singapore: World Scientific, 1996) [hep-th/9506084].

[72] I. Hinchliffe (for the Particle Data Group (PDG)), *J. Phys. G: Nucl. Part. Phys.* **33**, 110 (2006).

[73] G. R. Dvali and Z. Kakushadze, *Nucl. Phys. B* **537**, 297 (1999) [hep-th/9807140].

[74] G. Gabadadze and M. Shifman, *Phys. Rev. D* **61**, 075014 (2000) [hep-th/9910050].

[75] A. Armoni and M. Shifman, *Nucl. Phys. B* **664**, 233 (2003) [hep-th/0304127].

[76] A. Kovner, M. A. Shifman, and A. Smilga, *Phys. Rev. D* **56**, 7978 (1997) [hep-th/9706089].

[77] A. Ritz, M. Shifman, and A. Vainshtein, *Phys. Rev. D* **66**, 065015 (2002) [hep-th/0205083].

[78] A. Ritz, M. Shifman, and A. Vainshtein, *Phys. Rev. D* **70**, 095003 (2004) [hep-th/0405175].

[79] G. W. Gibbons and P. K. Townsend, *Phys. Rev. Lett.* **83**, 1727 (1999) [hep-th/9905196].

[80] S. M. Carroll, S. Hellerman, and M. Trodden, *Phys. Rev. D* **61**, 065001 (2000) [hep-th/9905217].

[81] H. Oda, K. Ito, M. Naganuma and N. Sakai, *Phys. Lett. B* **471**, 140 (1999) [hep-th/9910095].

[82] D. Binosi and T. ter Veldhuis, *Phys. Lett. B* **476**, 124 (2000) [hep-th/9912081].

[83] M. Shifman and T. ter Veldhuis, *Phys. Rev. D* **62**, 065004 (2000) [hep-th/9912162].

[84] J. P. Gauntlett, D. Tong, and P. K. Townsend, *Phys. Rev. D* **63**, 085001 (2001) [hep-th/0007124].

[85] K. Kakimoto and N. Sakai, *Phys. Rev. D* **68**, 065005 (2003) [hep-th/0306077].

[86] M. Eto, Y. Isozumi, M. Nitta, K. Ohashi, and N. Sakai, *Phys. Rev. D* **72**, 085004 (2005) [hep-th/0506135].

[87] M. Eto, Y. Isozumi, M. Nitta, K. Ohashi, and N. Sakai, *Phys. Lett. B* **632**, 384 (2006) [hep-th/0508241].

[88] M. Eto, T. Fujimori, T. Nagashima, M. Nitta, K. Ohashi, and N. Sakai, *Phys. Rev. D* **75**, 045010 (2007) [hep-th/0612003].

[89] M. Eto, T. Fujimori, T. Nagashima, M. Nitta, K. Ohashi, and N. Sakai, *Phys. Rev. D* **76**, 125025 (2007) [arXiv:0707.3267 [hep-th]].

[90] C. H. Taubes, *Commun. Math. Phys.* **72**, 277 (1980).

[91] E. R. Bezerra de Mello, *Mod. Phys. Lett. A* **5**, 581 (1990).

[92] J. R. Schmidt, *Phys. Rev. D* **46**, 1839 (1992).

[93] J. D. Edelstein, C. Nuñez, and F. Schaposnik, *Phys. Lett. B* **329**, 39 (1994) [hep-th/9311055].

[94] A. N. Redlich, *Phys. Rev. Lett.* **52**, 18 (1984); *Phys. Rev. D* **29**, 2366 (1984).

[95] L. Alvarez-Gaumè and E. Witten, *Nucl. Phys. B* **234**, 269 (1984).

[96] O. Aharony, A. Hanany, K. A. Intriligator, N. Seiberg, and M. J. Strassler, *Nucl. Phys. B* **499**, 67 (1997) [hep-th/9703110].

[97] H. Nishino and S. J. J. Gates, *Int. J. Mod. Phys. A* **8**, 3371 (1993).

[98] J. de Boer, K. Hori, and Y. Oz, *Nucl. Phys. B* **500**, 163 (1997) [hep-th/9703100].

[99] A. Rebhan, P. van Nieuwenhuizen, and R. Wimmer, *Nucl. Phys. B* **679**, 382 (2004) [hep-th/0307282].

[100] D. V. Vassilevich, *Phys. Rev. D* **68**, 045005 (2003) [hep-th/0304267].

[101] S. Ölmez and M. Shifman, arXiv:0808.1859 [hep-th].

[102] A. A. Penin, V. A. Rubakov, P. G. Tinyakov, and S. V. Troitsky, *Phys. Lett. B* **389**, 13 (1996) [hep-ph/9609257].

[103] A. Yung, *Nucl. Phys. B* **562**, 191 (1999) [hep-th/9906243].

[104] A. Gorsky, M. Shifman, and A. Yung, *Phys. Rev. D* **75**, 065032 (2007) [hep-th/0701040].

[105] G. 't Hooft, *Nucl. Phys. B* **79**, 276 (1974).

[106] A. M. Polyakov, Pisma Zh. *Eksp. Teor. Fiz.* **20**, 430 (1974) [Engl. transl. *JETP Lett.* **20**, 194 (1974), reprinted in *Solitons and Particles*, eds. C. Rebbi and G. Soliani (Singapore: World Scientific, 1984), p. 522].

[107] H. Georgi and S. L. Glashow, *Phys. Rev. Lett.* **28**, 1494 (1972).

[108] P. A. M. Dirac, *Proc. Roy. Soc, A* **133**, 60 (1931).

[109] E. Mottola, *Phys. Lett. B* **79**, 242 (1978) (E) *B* **80**, 433 (1979).

[110] See e.g. S. R. Coleman, The magnetic monopole: fifty years later, in *The Unity of the Fundamental Interactions, Proceedings of the 1981 International School of Subnuclear Physics*, Erice, Italy, ed. A. Zichichi (New York: Plenum Press, 1983).

[111] P. Goddard, J. Nuyts, and D. Olive, *Nucl. Phys. B* **125**, 1 (1977).

[112] E. J. Weinberg, *Nucl. Phys. B* **167**, 500 (1980); *Nucl. Phys. B* **203**, 445 (1982).

[113] E. Witten, *Phys. Lett. B* **86**, 283 (1979).

[114] R. K. Kaul, *Phys. Lett. B* **143**, 427 (1984).

[115] V. A. Rubakov, *Classical Theory of Gauge Fields* (Princeton University Press, 2002).

[116] C. Callias, *Commun. Math. Phys.* **62**, 213 (1978).

[117] R. Jackiw and C. Rebbi, *Phys. Rev. D* **13**, 3398 (1976) [reprinted in *Solitons and Particles*, eds. C. Rebbi and G. Soliani (Singapore: World Scientific, 1984), p. 331].

[118] C. Montonen and D. I. Olive, *Phys. Lett. B* **72**, 117 (1977).

[119] C. Mayrhofer, A. Rebhan, P. van Nieuwenhuizen, and R. Wimmer, *JHEP* **0709**, 069 (2007) [arXiv:0706.4476 [hep-th]].

[120] K. Hori and C. Vafa, *Mirror symmetry,* hep-th/0002222.

[121] S. Cecotti, P. Fendley, K. A. Intriligator, and C. Vafa, *Nucl. Phys. B* **386**, 405 (1992) [hep-th/9204102]; P. Fendley and K. A. Intriligator, *Nucl. Phys. B* **372**, 533 (1992) [hep-th/9111014]; S. Cecotti and C. Vafa, *Commun. Math. Phys.* **158**, 569 (1993) [hep-th/9211097].

[122] S. Ölmez and M. Shifman, *J. Phys. A* **40**, 11151 (2007) [hep-th/0703149].

[123] E. Witten, *Nucl. Phys. B* **202**, 253 (1982).

[124] G. 't Hooft, *Nucl. Phys. B* **190**, 455 (1981).

[125] S. Mandelstam, *Phys. Rept.* **23**, 245 (1976).

[126] M. R. Douglas and S. H. Shenker, *Nucl. Phys. B* **447**, 271 (1995) [hep-th/9503163].

[127] A. Hanany, M. J. Strassler, and A. Zaffaroni, *Nucl. Phys. B* **513**, 87 (1998) [hep-th/9707244].

[128] M. Strassler, *Prog. Theor. Phys. Suppl.* **131**, 439 (1998) [hep-lat/9803009].

[129] A. Yung, *What Do We Learn About Confinement From The Seiberg–Witten Theory?*, Proc. of 28th PNPI Winter School of Physics, St. Petersburg, Russia, 2000 [hep-th/0005088].

[130] A. Hanany and D. Tong, *JHEP* **0307**, 037 (2003) [hep-th/0306150].

[131] R. Auzzi, S. Bolognesi, J. Evslin, K. Konishi, and A. Yung, *Nucl. Phys. B* **673**, 187 (2003) [hep-th/0307287].

[132] M. Shifman and A. Yung, *Phys. Rev. D* **70**, 045004 (2004) [hep-th/0403149].

[133] A. Hanany and D. Tong, *JHEP* **0404**, 066 (2004) [hep-th/0403158].

[134] H. J. de Vega and F. A. Schaposnik, *Phys. Rev. Lett.* **56**, 2564 (1986); *Phys. Rev. D* **34**, 3206 (1986).

[135] J. Heo and T. Vachaspati, *Phys. Rev. D* **58**, 065011 (1998) [hep-ph/9801455].

[136] P. Suranyi, *Phys. Lett. B* **481**, 136 (2000) [hep-lat/9912023].

[137] F. A. Schaposnik and P. Suranyi, *Phys. Rev. D* **62**, 125002 (2000) [hep-th/0005109].

[138] M. A. C. Kneipp and P. Brockill, *Phys. Rev. D* **64**, 125012 (2001) [hep-th/0104171].

[139] K. Konishi and L. Spanu, *Int. J. Mod. Phys. A* **18**, 249 (2003) [hep-th/0106175].

[140] A. Marshakov and A. Yung, *Nucl. Phys. B* **647**, 3 (2002) [hep-th/0202172].

[141] M. Shifman and A. Yung, *Phys. Rev. D* **77**, 066008 (2008) [arXiv:0712.3512 [hep-th]].

[142] M. Shifman and A. Yung, *Phys. Rev. D* **67**, 125007 (2003) [hep-th/0212293].

[143] P. C. Argyres, M. R. Plesser, and N. Seiberg, *Nucl. Phys. B* **471**, 159 (1996) [hep-th/9603042].

[144] G. Carlino, K. Konishi, and H. Murayama, *Nucl. Phys. B* **590**, 37 (2000) [hep-th/0005076].

[145] P. Fayet and J. Iliopoulos, *Phys. Lett. B* **51**, 461 (1974).

[146] K. Bardakci and M. B. Halpern, *Phys. Rev. D* **6**, 696 (1972).

[147] A. Achucarro and T. Vachaspati, *Phys. Rept.* **327**, 427 (2000) [hep-ph/9904229].

[148] W. Garcia Fuertes and J. Mateos Guilarte, *Phys. Lett. B* **437**, 82 (1998)
       [hep-th/9807218]. J. D. Edelstein, W. G. Fuertes, J. Mas and J. M. Guilarte, *Phys.
       Rev. D* **62**, 065008 (2000) [hep-th/0001184].

[149] E. Witten, *Nucl. Phys. B* **249**, 557 (1985).

[150] M. Hindmarsh, *Phys. Lett. B* **225**, 127 (1989).

[151] M. G. Alford, K. Benson, S. R. Coleman, J. March-Russell, and F. Wilczek, *Nucl.
       Phys. B* **349**, 414 (1991).

[152] M. Eto, M. Nitta, and N. Sakai, *Nucl. Phys. B* **701**, 247 (2004) [hep-th/0405161].

[153] S. Bolognesi, *Nucl. Phys. B* **719**, 67 (2005) [hep-th/0412241].

[154] A. Gorsky, M. Shifman, and A. Yung, *Phys. Rev. D* **71**, 045010 (2005)
       [hep-th/0412082].

[155] V. Markov, A. Marshakov, and A. Yung, *Nucl. Phys. B* **709**, 267 (2005)
       [hep-th/0408235].

[156] V. A. Novikov, M. A. Shifman, A. I. Vainshtein, and V. I. Zakharov, *Phys. Reports*
       116, 103 (1984).

[157] E. Witten, *Nucl. Phys. B* **403**, 159 (1993) [hep-th/9301042].

[158] A. M. Polyakov, *Phys. Lett. B* **59**, 79 (1975).

[159] E. Witten, *Nucl. Phys. B* **149**, 285 (1979).

[160] S. R. Coleman, *Commun. Math. Phys.* **31**, 259 (1973).

[161] V. A. Fateev, I. V. Frolov, and A. S. Schwarz, *Sov. J. Nucl. Phys.* **30**, 590 (1979)
       [Yad. Fiz. **30**, 1134 (1979)]; *Nucl. Phys. B* **154** (1979) 1. See also in A. Polyakov,
       Gauge Fields and Strings (New York: Harwood Press, 1987).

[162] P. Fendley and K. A. Intriligator, *Nucl. Phys. B* **380**, 265 (1992) [hep-th/9202011].

[163] T. Eguchi, K. Hori and C. S. Xiong, *Int. J. Mod. Phys. A* **12**, 1743 (1997)
       [hep-th/9605225].

[164] A. Gorsky, M. Shifman, and A. Yung, *Phys. Rev. D* **73**, 065011 (2006)
       [hep-th/0512153].

[165] D. Tong, *Phys. Rev. D* **69**, 065003 (2004) [hep-th/0307302].

[166] F. A. Bais, *Phys. Lett. B* **98**, 437 (1981).

[167] M. Hindmarsh and T. W. B. Kibble, *Phys. Rev. Lett.* **55**, 2398 (1985).

[168] A. E. Everett and M. Aryal, *Phys. Rev. Lett.* **57**, 646 (1986).

[169] J. Preskill and A. Vilenkin, *Phys. Rev. D* **47**, 2324 (1993) [hep-ph/9209210].

[170] M. A. C. Kneipp, *Phys. Rev. D* **68**, 045009 (2003) [hep-th/0211049]; *Phys. Rev. D*
       69, 045007 (2004) [hep-th/0308086].

[171] R. Auzzi, S. Bolognesi, J. Evslin, and K. Konishi, *Nucl. Phys. B* **686**, 119 (2004)
       [hep-th/0312233]; R. Auzzi, S. Bolognesi, J. Evslin, K. Konishi, and H. Murayama,
       Nucl. Phys. B **701**, 207 (2004) [hep-th/0405070]; R. Auzzi, S. Bolognesi, and
       J. Evslin, *JHEP* **0502**, 046 (2005) [hep-th/0411074]; M. Eto, L. Ferretti,
       K. Konishi, G. Marmorini, M. Nitta, K. Ohashi, W. Vinci, and N. Yokoi, *Nucl. Phys.
       B* **780**, 181 (2007) [hep-th/0611313].

[172] Y. Isozumi, M. Nitta, K. Ohashi, and N. Sakai, *Phys. Rev. D* **71** 065018 (2005)
       [hep-th/0405129].

[173] D. Tong, *JHEP* **0612**, 051 (2006) [hep-th/0610214].

[174] P. C. Argyres and M. R. Douglas, *Nucl. Phys. B* **448**, 93 (1995) [hep-th/9505062].

[175] K. Evlampiev and A. Yung, *Nucl. Phys. B* **662**, 120 (2003) [hep-th/0303047].

[176] A. M. Polyakov and A. A. Belavin, *JETP Lett.* **22**, 245 (1975).

[177] R. S. Ward, *Phys. Lett. B* **158**, 424 (1985).

[178] R. A. Leese and T. M. Samols, *Nucl. Phys. B* **396**, 639 (1993).

[179] M. Shifman and A. Yung, *Phys. Rev. D* **73**, 125012 (2006) [hep-th/0603134].

[180] M. Eto, J. Evslin, K. Konishi, G. Marmorini, M. Nitta, K. Ohashi, V. Vinci, and N. Yakoi, *Phys. Rev. D* **76**, 105002 (2007) [arXiv:0704.2218 (hep-th)].

[181] N. Dorey, T. J. Hollowood, and D. Tong, *JHEP* **9905**, 006 (1999) [hep-th/9902134].

[182] R. A. Leese, *Phys. Rev. D* **46**, 4677 (1992).

[183] R. Auzzi, M. Shifman, and A. Yung, *Phys. Rev. D* **73**, 105012 (2006) [hep-th/0511150].

[184] K. Hashimoto and D. Tong, *JCAP* **0509**, 004 (2005) [hep-th/0506022].

[185] M. Eto, K. Konishi, G. Marmorini, M. Nitta, K. Ohashi, W. Vinci, and N. Yokoi, *Phys. Rev. D* **74** (2006) 065021 [hep-th/0607070].

[186] M. Eto, K. Hashimoto, G. Marmorini, M. Nitta, K. Ohashi, and W. Vinci, *Phys. Rev. Lett.* **98**, 091602 (2007) [hep-th/0609214].

[187] N. S. Manton and J. M. Speight, *Commun. Math. Phys.* **236**, 535 (2003) [hep-th/0205307].

[188] M. Eto, Y. Isozumi, M. Nitta, K. Ohashi, and N. Sakai, *Phys. Rev. Lett.* **96**, 161601 (2006) [hep-th/0511088].

[189] M. Shifman and A. Yung, *Phys. Rev. D* **72**, 085017 (2005) [hep-th/0501211].

[190] M. Edalati and D. Tong, *JHEP* **0705**, 005 (2007) [hep-th/0703045].

[191] M. Shifman and A. Yung, *Phys. Rev. D* **77**, 125016 (2008), arXiv:0803.0158.

[192] M. Shifman and A. Yung, *Phys. Rev. D* **77**, 125017 (2008), arXiv:0803.0698.

[193] D. Tong, *JHEP* **0709**, 022 (2007) [hep-th/0703235].

[194] G. Veneziano and S. Yankielowicz, *Phys. Lett. B* **113**, 231 (1982).

[195] E. Witten, *JHEP* **9707** 003 (1997) [arXiv:hep-th/9707093].

[196] N. Seiberg, *Nucl. Phys. B* **435**, 129 (1995) [hep-th/9411149].

[197] K. A. Intriligator and N. Seiberg, *Nucl. Phys. Proc. Suppl.* **45BC**, 1 (1996) [hep-th/9509066]; *Lectures on Supersymmetry Breaking*, hep-ph/0702069.

[198] M. Shifman, *Prog. Part. Nucl. Phys.* **39**, 1 (1997) [hep-th/9704114].

[199] K. Intriligator, N. Seiberg, and D. Shih, *JHEP* **0604**, 021 (2006) [hep-th/0602239].

[200] M. Eto, K. Hashimoto, and S. Terashima, *JHEP* **0703**, 061 (2007) [arXiv:hep-th/0610042].

[201] J. E. Kiskis, *Phys. Rev. D* **15**, 2329 (1977); M. M. Ansourian, *Phys. Lett. B* **70**, 301 (1977); N. K. Nielsen and B. Schroer, *Nucl. Phys. B* **120**, 62 (1977); R. Jackiw and P. Rossi, *Nucl. Phys. B* **190**, 681 (1981); E. J. Weinberg, *Phys. Rev. D* **24**, 2669 (1981).

[202] M. A. Shifman, *Phys. Reports* **209**, 341 (1991).

[203] S. Coleman, The uses of instantons, in S. Coleman, *Aspects of Symmetry* (Cambridge: Cambridge University Press, 1985), p. 265.

[204] A. B. Zamolodchikov and Al. B. Zamolodchikov, *Annals Phys.* **120**, 253 (1979).

[205] A. B. Zamolodchikov and Al. B. Zamolodchikov, *Nucl. Phys. B* **379**, 602 (1992).

[206] S. R. Coleman, *Annals Phys.* **101**, 239 (1976).

[207] E. Witten, *Phys. Rev. Lett.* **81**, 2862 (1998) [hep-th/9807109].

[208] M. Shifman, *Phys. Rev. D* **59**, 021501 (1999) [hep-th/9809184].

[209] C. Vafa and E. Witten, *Nucl. Phys. B* **431**, 3 (1994) [hep-th/9408074].

[210] R. Donagi and E. Witten, *Nucl. Phys. B* **460**, 299 (1996) [hep-th/9510101].

[211] A. Achucarro, A. C. Davis, M. Pickles, and J. Urrestilla, *Phys. Rev. D* **66**, 105013 (2002) [hep-th/0109097].

[212] M. Pickles and J. Urrestilla, *JHEP* **0301**, 052 (2003) [hep-th/0211240].

[213] I. I. Kogan, A. Kovner, and M. A. Shifman, *Phys. Rev. D* **57**, 5195 (1998) [hep-th/9712046].

[214] E. R. Abraham, P. K. Townsend, *Phys. Lett. B* **291**, 85 (1992); *Phys. Lett. B* **295**, 225 (1992).

[215] J. P. Gauntlett, R. Portugues, D. Tong, and P. K. Townsend, *Phys. Rev. D* **63**, 085002 (2001) [hep-th/0008221].

[216] S. L. Dubovsky and V. A. Rubakov, *Int. J. Mod. Phys. A* **16**, 4331 (2001) [hep-th/0105243].

[217] G. Dvali and A. Vilenkin, *Phys. Rev. D* **67**, 046002 (2003) [hep-th/0209217]; G. Dvali, H. B. Nielsen, and N. Tetradis, *Phys. Rev. D* **77**, 085005 (2008) [arXiv:0710.5051 [hep-th]]; R. Auzzi, S. Bolognesi, M. Shifman, and A. Yung, *Confinement and Localization on Domain Walls*, arXiv:0807.1908 [hep-th].

[218] N. Sakai and D. Tong, *JHEP* **0503**, 019 (2005) [hep-th/0501207].

[219] M. Eto, T. Fujimori, M. Nitta, K. Ohashi, and N. Sakai, "Domain walls with non-Abelian clouds", hep-th/0802.3135.

[220] J. P. Gauntlett, D. Tong, and P. K. Townsend, *Phys. Rev. D* **64**, 025010 (2001) [hep-th/0012178].

[221] D. Tong, *Phys. Rev. D* **66**, 025013 (2002) [hep-th/0202012].

[222] Y. Isozumi, M. Nitta, K. Ohashi, and N. Sakai, *Phys. Rev. Lett.* **93**, 161601 (2004) [hep-th/0404198].

[223] Y. Isozumi, M. Nitta, K. Ohashi, and N. Sakai, *Phys. Rev. D* **70**, 125014 (2004) [hep-th/0405194].

[224] M. Eto, Y. Isozumi, M. Nitta, K. Ohashi, K. Ohta, N. Sakai, and Y. Tachikawa, *Phys. Rev. D* **71**, 105009 (2005) [hep-th/0503033]; N. Sakai and Y. Yang, Commun. Math. Phys. **267**, 783 (2006) [hep-th/0505136].

[225] M. Arai, M. Naganuma, M. Nitta, and N. Sakai, *Nucl. Phys. B* **652**, 35 (2003) [hep-th/0211103]; BPS wall in $\mathcal{N} = 2$ SUSY nonlinear sigma model with Eguchi–Hanson manifold, in *Garden of Quanta*, ed. A. Arafune *et al.* (Singapore: World Scientific, 2003) p. 299 [hep-th/0302028].

[226] R. Auzzi, M. Shifman, and A. Yung, *Phys. Rev. D* **74**, 045007 (2006) [hep-th/0606060].

[227] A. M. Polyakov, *Nucl. Phys. B* **120**, 429 (1977).

[228] M. Shifman and A. Yung, *Phys. Rev. D* **74**, 045006 (2006) [hep-th/0603236].

[229] N. D. Lambert and D. Tong, *Nucl. Phys. B* **569**, 606 (2000) [hep-th/9907098].

[230] Y. Isozumi, K. Ohashi, and N. Sakai, *JHEP* **0311**, 060 (2003) [hep-th/0310189].

[231] R. Auzzi, M. Shifman, and A. Yung, *Phys. Rev. D* **72**, 025002 (2005) [hep-th/0504148].

[232] D. Tong, *JHEP* **0602**, 030 (2006) [hep-th/0512192].

[233] K. Intriligator and N. Seiberg, *Phys. Lett. B* **387**, 513 (1996) [hep-th/9607207].

[234] J. de Boer, K. Hori, H. Ooguri, and Y. Oz, *Nucl. Phys. B* **493**, 101 (1997) [hep-th/9611063].

[235] A. Armoni and T. J. Hollowood, *JHEP* **0507**, 043 (2005) [hep-th/0505213]; JHEP **0602**, 072 (2006) [hep-th/0601150].

[236] V. A. Novikov, M. A. Shifman, A. I. Vainshtein, and V. I. Zakharov, *Nucl. Phys. B* **229**, 381 (1983); *Phys. Lett. B* **166**, 329 (1986); M. A. Shifman, A. I. Vainshtein and V. I. Zakharov, *Sov. J. Nucl. Phys.* **43**, 1028 (1986); *Phys. Lett. B* **166**, 334 (1986); M. A. Shifman and A. I. Vainshtein, *Sov. J. Nucl. Phys.* **44**, 321 (1986).

[237] E. Witten, *Phys. Rev. D* **16**, 2991 (1977).

[238] R. Rajaraman, *Solitons And Instantons* (Amsterdam: North-Holland, 1982).

[239] H. Bohr, E. Katznelson, and K. S. Narain, *Nucl. Phys. B* **238**, 407 (1984).

[240] A. M. Perelomov, *Phys. Rept.* **146**, 135 (1987); *Phys. Rept.* **174**, 229 (1989).

[241] B. Zumino, *Phys. Lett. B* **87**, 203 (1979).

[242] H. Eichenherr, *Nucl. Phys. B* **146**, 215 (1978) [Erratum-ibid. B **155**, 544 (1979)]; V. L. Golo and A. M. Perelomov, *Lett. Math. Phys.* **2**, 477 (1978); E. Cremmer and J. Scherk, *Phys. Lett. B* **74**, 341 (1978).

Index

Author index

Printed in the United States
by Baker & Taylor Publisher Services